Building Python Web APIs with FastAPI

A fast-paced guide to building high-performance, robust web APIs with very little boilerplate code

Abdulazeez Abdulazeez Adeshina

BIRMINGHAM—MUMBAI

Building Python Web APIs with FastAPI

Associate Group Product Manager: Pavan Ramchandani

Publishing Product Manager: Aaron Tanna

Senior Editor: Mark Dsouza

Content Development Editor: Divya Vijayan

Technical Editor: Shubham Sharma

Copy Editor: Safis Editing

Project Coordinator: Rashika Ba

Proofreader: Safis Editing

Indexer: Pratik Shirodkar

Production Designer: Vijay Kamble

Marketing Coordinators: Anamika Singh and Marylou De Mello

First published: July 2022

Production reference: 1150722

Published by Packt Publishing Ltd.

Livery Place

35 Livery Street

Birmingham

B3 2PB, UK.

ISBN 978-1-80107-663-0

www.packt.com

To my mother, and to the memory of my father, for their sacrifices, belief, and constant support throughout the years. To my amazing sisters, Ameedat and Aminat, for being a continuous source of joy and happiness. To my uncle, Bako, for his continuous support over the years. To my best friends, Abdulrahman and Amina, for always being there.

– Abdulazeez Abdulazeez Adeshina

Contributors

About the author

Abdulazeez Abdulazeez Adeshina is a skilled Python developer, backend software engineer, and technical writer, with a wide range of technical skill sets in his arsenal. His background has led him to build command-line applications, backend applications in FastAPI, and algorithm-based treasure-hunting tools. He also enjoys teaching Python and solving mathematical-oriented problems through his blog. Abdulazeez is currently in his penultimate year of a water resources and environmental engineering program. His work experience as a guest technical author includes the likes of Auth0, LogRocket, Okteto, and TestDriven.

I want to thank Allah (SWT) for his unending grace, and Sebastián Ramírez for creating FastAPI. I want to thank Precious Ndubueze for introducing me to FastAPI and insisting I get acquainted with the framework, and Bolaji Olajide for helping me by reviewing the first chapters. Lastly, I also want to thank every one of my close associates, especially my uncle, Tosin Olayanju, for their support throughout the development of this book – I am incredibly grateful and honored.

About the reviewer

Akash Ranjan is a Python professional, with 6+ years of industry experience. He has developed, deployed, and managed large-scale enterprise applications. He has extensive experience in building APIs and designing microservice-based application architecture.

Table of Contents

3

Response Models and Error Handling

4

Templating in FastAPI

Part 2: Building and Securing FastAPI Applications

5

Structuring FastAPI Applications

6
Connecting to a Database

7
Securing FastAPI Applications

Part 3: Testing And Deploying FastAPI Applications

8
Testing FastAPI Applications

9
Deploying FastAPI Applications

Index

Other Books You May Enjoy

Preface

FastAPI is a fast and efficient web framework for building APIs with Python. This book is a thorough guide on how to build an application with the FastAPI framework.

It starts with the basics of the FastAPI framework and the other technologies used throughout this book. You'll then learn about the different aspects of the framework: the routing system, response modeling, error handling, and templating.

In this book, you will learn how to build fast, efficient, and scalable applications in **Python** with **FastAPI**. You will begin from a *Hello World* application to a complete API that uses a database, authentication, and templates. You will learn how to structure your application to enhance efficiency, readability, and scalability. By integrating with other libraries in your application, you will learn how to connect your application to both a **SQL database** and a **NoSQL database**, integrate templates, and build authentication. Toward the end of this book, you will learn how to write tests, containerize your application, build a continuous integration and delivery pipeline using GitHub actions, and deploy your application to multiple cloud services. All of these will be taught via a theoretical and practical approach.

By the end of this book, you'll be equipped with the necessary knowledge to build and deploy a robust web API using the FastAPI framework.

Who this book is for

The primary audience for this book is any Python developer interested in building web APIs. The ideal reader is familiar with the basics of the Python programming language.

What this book covers

Chapter 1, *Getting Started with FastAPI*, introduces the basics of FastAPI and other technologies used in the book. The chapter also details the steps involved in setting up your development environment for your FastAPI application.

Chapter 2, Routing in FastAPI, talks in detail about the process of creating endpoints using the routing system in FastAPI. The components of a routing system, including the request body and path parameters, are also discussed alongside validating them with pydantic classes.

Chapter 3, Response Models and Error Handling, introduces responses in FastAPI, response modeling, error handling, and status codes.

Chapter 4, Templating in FastAPI, discusses how templates can be used to create views and render responses from the API.

Chapter 5, Structuring FastAPI Applications, introduces structuring applications and also briefly introduces the application to be built in the next chapters.

Chapter 6, Connecting to a Database, discusses two classes of databases (SQL and NoSQL) and demonstrates how you can connect your FastAPI application to either of them. We'll cover how to connect to and use a SQL database using SQLModel, and how to work with MongoDB using an object-document mapper, Beanie.

Chapter 7, Securing FastAPI Applications, talks about what securing your application entails – authorization and authentication, implementing authentication, and restricting access to application endpoints.

Chapter 8, Testing FastAPI Applications, explains what testing is and how to test our API endpoints.

Chapter 9, Deploying FastAPI Applications, discusses the steps involved in deploying your FastAPI application.

To get the most out of this book

You will need the latest version of Python installed on your computer. You also need to be familiar with the Python programming language to get the most out of this book.

Software/hardware covered in the book	Operating system requirements
Python 3.10	Windows, macOS, or Linux
Git 2.36.0	Windows, macOS, or Linux

If you are using the digital version of this book, we advise you to type the code yourself or access the code from the book's GitHub repository (a link is available in the next section). Doing so will help you avoid any potential errors related to the copying and pasting of code.

Download the example code files

You can download the example code files for this book from GitHub at `https://github.com/PacktPublishing/Building-Python-Web-APIs-with-FastAPI`. If there's an update to the code, it will be updated in the GitHub repository.

We also have other code bundles from our rich catalog of books and videos available at `https://github.com/PacktPublishing/`. Check them out!

Download the color images

We also provide a PDF file that has color images of the screenshots and diagrams used in this book. You can download it here: `https://packt.link/qqhpc`.

Conventions used

There are a number of text conventions used throughout this book.

`Code in text`: Indicates code words in text, database table names, folder names, filenames, file extensions, pathnames, dummy URLs, user input, and Twitter handles. Here is an example: "To switch back to the original `main` branch, we run `git checkout main`."

A block of code is set as follows:

```
from fastapi import FastAPI
from routes.user import user_router

import uvicorn
```

When we wish to draw your attention to a particular part of a code block, the relevant lines or items are set in bold:

```
from pydantic import BaseModel
from typing import List

class Event(BaseModel):
    id: int
    title: str
    image: str
    description: str
```

```
    tags: List[str]
    location: str
```

Any command-line input or output is written as follows:

```
$ git add hello.txt
$ git commit -m "Initial commit"
```

Bold: Indicates a new term, an important word, or words that you see onscreen. For instance, words in menus or dialog boxes appear in **bold**. Here is an example: "As shown in the previous model diagram, each user will have an **Events** field, which is a list of the events they have ownership of."

> Tips or Important Notes
> Appear like this.

Get in touch

Feedback from our readers is always welcome.

General feedback: If you have questions about any aspect of this book, email us at customercare@packtpub.com and mention the book title in the subject of your message.

Errata: Although we have taken every care to ensure the accuracy of our content, mistakes do happen. If you have found a mistake in this book, we would be grateful if you would report this to us. Please visit www.packtpub.com/support/errata and fill in the form.

Piracy: If you come across any illegal copies of our works in any form on the internet, we would be grateful if you would provide us with the location address or website name. Please contact us at copyright@packt.com with a link to the material.

If you are interested in becoming an author: If there is a topic that you have expertise in and you are interested in either writing or contributing to a book, please visit authors.packtpub.com.

Share Your Thoughts

Once you've read *Building Python Web APIs with FastAPI*, we'd love to hear your thoughts! Scan the QR code below to go straight to the Amazon review page for this book and share your feedback.

https://packt.link/r/1801076634

Your review is important to us and the tech community and will help us make sure we're delivering excellent quality content.

Part 1: An Introduction to FastAPI

Upon completion of this part, you will have a substantial understanding of FastAPI, including routing, file handling, error handling, building response models, and templating. This part starts by introducing the technologies that will be used throughout the book before proceeding to the basics of the FastAPI framework.

This part comprises the following chapters:

- *Chapter 1, Getting Started with FastAPI*
- *Chapter 2, Routing in FastAPI*
- *Chapter 3, Response Models and Error Handling*
- *Chapter 4, Templating in FastAPI*

1
Getting Started with FastAPI

FastAPI is the Python web framework that we are going to use in this book. It is a fast, lightweight modern API and has an easier learning curve when compared to other Python-based web frameworks, such as Flask and Django. FastAPI is relatively new, but it has a growing community. It is used extensively in building web APIs and in deploying machine learning models.

In the first chapter, you will learn how to set up your development environment and build your first FastAPI application. You will begin by learning the basics of **Git** – a version control system – to equip you with the knowledge of storing, tracking, and retrieving file changes as you build your application. You will also learn how to handle packages in Python using pip, how to create isolated development environments with **Virtualenv**, and the basics of **Docker**. Lastly, you will be introduced to the basics of FastAPI by building a simple *Hello World* application.

An understanding of the technologies previously mentioned is required to build a full-blown FastAPI application. It also serves as an addition to your current skillset.

At the completion of this chapter, you will be able to set up and use Git, install and manage packages using pip, create an isolated development environment with Virtualenv, use Docker, and most importantly, scaffold a FastAPI application.

This chapter covers the following topics:

- Git basics
- Creating isolated development environments with Virtualenv
- Package management with pip
- Setting up and learning the basics of Docker
- Building a simple FastAPI application

Technical Requirement

You can find the code files for this chapter on GitHub at `https://github.com/PacktPublishing/Building-Python-Web-APIs-with-FastAPI/tree/main/ch01`

Git basics

Git is a version control system that enables developers to record, keep track, and revert to earlier versions of files. It is a decentralized and lightweight tool that can be installed on any operating system.

You will be learning how to use Git for record-keeping purposes. As each layer of the application is being built, changes will be made, and it's important that these changes are kept note of.

Installing Git

To install Git, visit the downloads page at `https://git-scm.com/downloads` and select a download option for your current operating system. You'll be redirected to an instructional page on how to install Git on your machine.

It is also worth noting that Git comes as a **CLI** and a **GUI** application. Therefore, you can download the one that works best for you.

Git operations

As mentioned earlier, Git can be used to record, track, and revert to earlier versions of a file. However, only the basic operations of Git will be used in this book and will be introduced in this section.

In order for Git to run properly, folders housing files must be initialized. Initializing folders enables Git to keep track of the content except otherwise exempted.

To initialize a new Git repository in your project, you need to run the following command in your terminal:

```
$ git init
```

To enable tracking of files, a file must first be added and committed. A Git commit enables you to track file changes between timeframes; for example, a commit made an hour ago and the current file version.

> **What Is a Commit?**
>
> A commit is a unique capture of a file or folder status at a particular time, and it is identified by a unique code.

Now that we know what a commit is, we can go ahead and commit a file as follows:

```
$ git add hello.txt
$ git commit -m "Initial commit"
```

You can track the status of your files after making changes by running the following command:

```
$ git status
```

Your terminal should look similar to the following:

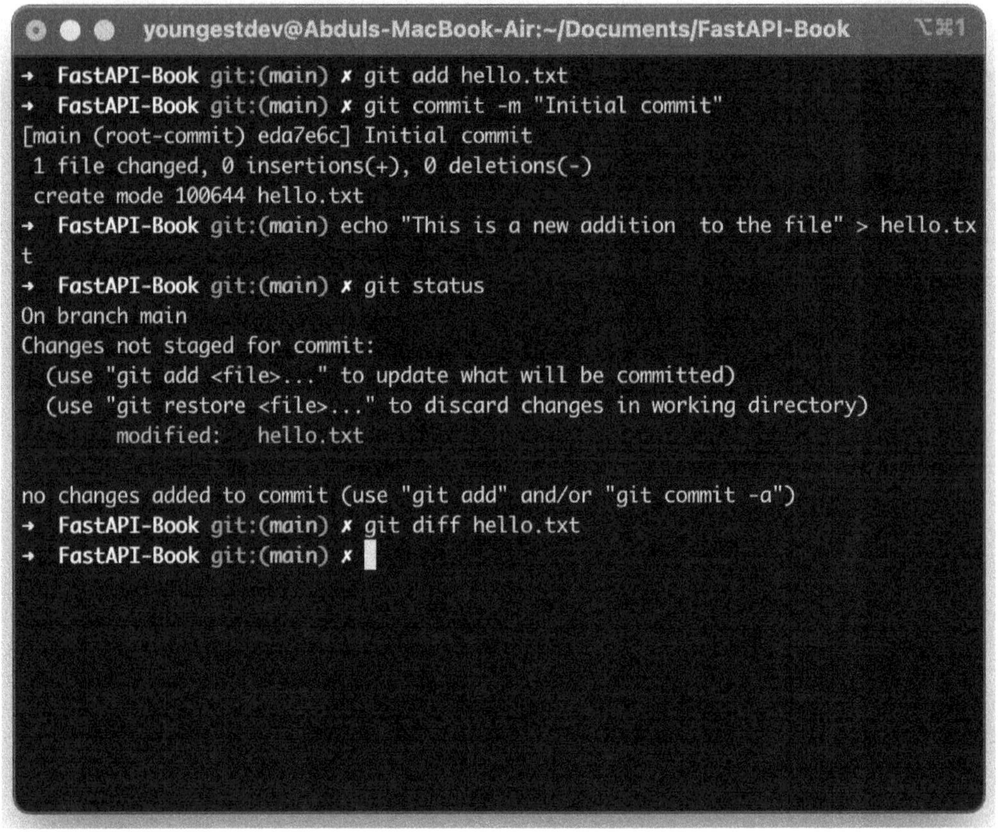

```
youngestdev@Abduls-MacBook-Air:~/Documents/FastAPI-Book

→  FastAPI-Book git:(main) ✗ git add hello.txt
→  FastAPI-Book git:(main) ✗ git commit -m "Initial commit"
[main (root-commit) eda7e6c] Initial commit
 1 file changed, 0 insertions(+), 0 deletions(-)
 create mode 100644 hello.txt
→  FastAPI-Book git:(main) echo "This is a new addition  to the file" > hello.tx
t
→  FastAPI-Book git:(main) ✗ git status
On branch main
Changes not staged for commit:
  (use "git add <file>..." to update what will be committed)
  (use "git restore <file>..." to discard changes in working directory)
        modified:   hello.txt

no changes added to commit (use "git add" and/or "git commit -a")
→  FastAPI-Book git:(main) ✗ git diff hello.txt
→  FastAPI-Book git:(main) ✗ ▌
```

Figure 1.1 – Git commands

To view the changes made to the file, which can be additions or subtractions from the file contents, run the following command:

```
$ git diff
```

Your terminal should look similar to the following:

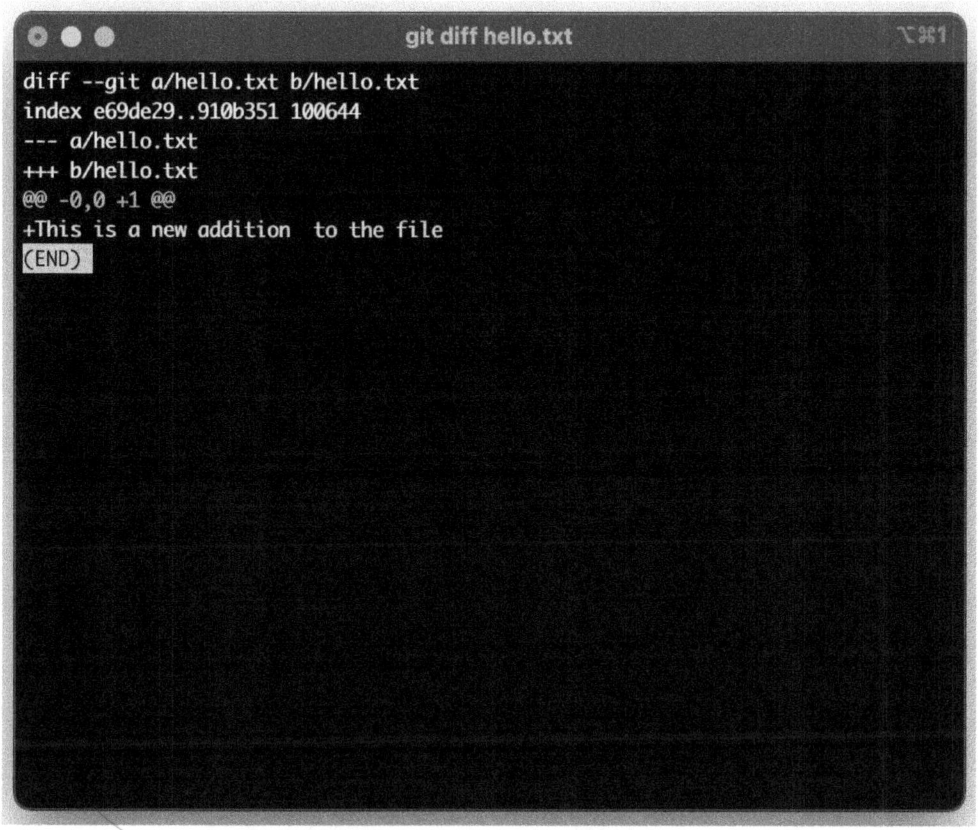

```
● ● ●                          git diff hello.txt                           ⌥⌘1
diff --git a/hello.txt b/hello.txt
index e69de29..910b351 100644
--- a/hello.txt
+++ b/hello.txt
@@ -0,0 +1 @@
+This is a new addition  to the file
(END)
```

Figure 1.2 – Output from the git diff command

It is good practice to include a `.gitignore` file in every folder. The `.gitignore` file contains the names of files and folders to be ignored by Git. This way, you can add and commit all the files in your folder without the fear of committing files like `.env`.

To include a `.gitignore` file, run the following command in your terminal:

```
$ touch .gitignore
```

To exempt a file from being tracked by Git, add it to the `.gitignore` file as follows:

```
$ echo ".env" >> .gitignore
```

Common files contained in a `.gitignore` file include the following:

- Environment files (`*.env`)
- Virtualenv folder (env, venv)
- IDE metadata folders (such as `.vscode` and `.idea`)

Git branches

Branches are an important feature that enables developers to easily work on different application features, bugs, and so on, separately before merging into the main branch. The system of branching is employed in both small-scale and large-scale applications and promotes the culture of previewing and collaborations via pull requests. The primary branch is called the main branch and it is the branch from which other branches are created.

To create a new branch from an existing branch, we run the `git checkout -b newbranch` command. Let's create a new branch by running the following command:

```
$ git checkout -b hello-python-branch
```

The preceding command creates a new branch from the existing one, and then sets the active branch to the newly created branch. To switch back to the original `main` branch, we run `git checkout main` as follows:

```
$ git checkout main
```

> **Important Note**
>
> Running `git checkout main` makes `main` the active working branch, whereas `git checkout -b newbranch` creates a new branch from the current working branch and sets the newly created branch as the active one.
>
> To learn more, refer to the Git documentation: `http://www.git-scm.com/doc`.

Now that we have learned the basics of Git, we can now proceed to learn about how to create isolated environments with **virtualenv**.

Creating isolated development environments with Virtualenv

The traditional approach to developing applications in Python is to isolate these applications in a virtual environment. This is done to avoid installing packages globally and reduce conflicts during application development.

A virtual environment is an isolated environment where application dependencies installed can only be accessed within it. As a result, the application can only access packages and interact only within this environment.

Creating a virtual environment

By default, the venv module from the standard library is installed in Python3. The venv module is responsible for creating a virtual environment. Let's create a todos folder and create a virtual environment in it by running the following commands:

```
$ mkdir todos && cd todos
$ python3 -m venv venv
```

The venv module takes an argument, which is the name of the folder where the virtual environment should be installed into. In our newly created virtual environment, a copy of the Python interpreter is installed in the lib folder, and the files enabling interactions within the virtual environment are stored in the bin folder.

Activating and deactivating the virtual environment

To activate a virtual environment, we run the following command:

```
$ source venv/bin/activate
```

The preceding command instructs your shell to use the virtual environment's interpreter and packages by default. Upon activating the virtual environment, a prefix of the venv virtual environment folder is added before the prompt as follows:

```
● ● ●    youngestdev@Abduls-MacBook-Air:~/Documents/FastAPI-Book        ⌥⌘1
→  FastAPI-Book git:(main) python3 -m pip list
Package                     Version
--------------------------  --------
cachetools                  4.2.2
certifi                     2021.5.30
chardet                     4.0.0
google-auth                 1.32.0
googleapis-common-protos    1.53.0
httplib2                    0.19.1
idna                        2.10
packaging                   20.9
pip                         21.1.2
protobuf                    3.17.3
pyasn1                      0.4.8
pyasn1-modules              0.2.8
pyparsing                   2.4.7
pytz                        2021.1
requests                    2.25.1
rsa                         4.7.2
setuptools                  41.2.0
six                         1.15.0
urllib3                     1.26.5
wheel                       0.33.1
WARNING: You are using pip version 21.1.2; however, version 21.2.4 is availab
le.
```

Figure 1.3 – Prefixed prompt

To deactivate a virtual environment, the deactivate command is run in the prompt. Running the command immediately exits the isolated environment and the prefix is removed as follows:

Figure 1.4 – Deactivating a virtual environment

> **Important Note**
> You can also create a virtual environment and manage application
> dependencies using *Pipenv* and *Poetry*.

Now that we have created the virtual environment, we can now proceed to understand
how package management with **pip** works.

Package management with pip

A FastAPI application constitutes packages, therefore you will be introduced to package
management practices, such as installing packages, removing packages, and updating
packages for your application.

Installing packages from the source can turn out to be a cumbersome task as, most of the time, it involves downloading and unzipping `.tar.gz` files before manual installation. In a scenario where a hundred packages are to be installed, this method becomes inefficient. Then, how do you automate this process?

Pip is a Python package manager like JavaScript's `yarn`; it enables you to automate the process of installing Python packages – both globally and locally.

Installing pip

Pip is automatically installed during a Python installation. You can verify whether pip is installed by running the following command in your terminal:

```
$ python3 -m pip list
```

The preceding command should return a list of packages installed. The output should be similar to the following figure:

Figure 1.5 – List of installed Python packages

If the command returns an error, follow the instructions at `https://pip.pypa.io/en/stable/installation/` to install pip.

Basic commands

With `pip` installed, let's learn its basic commands. To install the `FastAPI` package with pip, we run the following command:

```
$ pip install fastapi
```

On a Unix operating system, such as Mac or Linux, in some cases, the `sudo` keyword is prepended to install global packages.

To uninstall a package, the following command is used:

```
$ pip uninstall fastapi
```

To collate the current packages installed in a project into a file, we use the following `freeze` command:

```
$ pip freeze > requirements.txt
```

The > operator tells bash to save the output from the command into the `requirements.txt` file. This means that running `pip freeze` returns an output of all the currently installed packages.

To install packages from a file such as the `requirements.txt` file, the following command is used:

```
$ pip install -r requirements.txt
```

The preceding command is mostly used in deployment.

Now that you have learned the basics of pip and have gone over some basic commands, let's learn the basics of **Docker**.

Setting up Docker

As our application grows into having multiple layers, such as a database, coupling the application into a single piece enables us to deploy our application. We'll be using **Docker** to containerize our application layers into a single image, which can then be easily deployed locally or in the cloud.

Additionally, using a Dockerfile and a docker-compose file eliminates the need to upload and share images of our applications. New versions of our applications can be built from the Dockerfile and deployed using the docker-compose file. Application images can also be stored and retrieved from **Docker Hub**. This is known as a push and pull operation.

To begin setting up, download and install Docker from `https://docs.docker.com/ install`.

Dockerfile

A Dockerfile contains instructions on how our application image is to be built. The following is an example Dockerfile:

```
FROM PYTHON:3.8
# Set working directory to /usr/src/app
WORKDIR /usr/src/app
# Copy the contents of the current local directory into the
container's working directory
ADD . /usr/src/app
# Run a command
CMD ["python", "hello.py"]
```

Next, we'll build the application container image and tag it `getting_started` as follows:

```
$ docker build -t getting_started .
```

If the Dockerfile isn't present in the directory where the command is being run, the path to the Dockerfile should be properly appended as follows:

```
$ docker build -t api api/Dockerfile
```

The container image can be run using the following command:

```
$ docker run getting-started
```

Docker is an efficient tool for containerization. We have only looked at the basic operations and we'll learn more operations practically in *Chapter 9, Deploying FastAPI Applications*.

Building a simple FastAPI application

Finally, we can now get to our first FastAPI project. Our aim in this section is to introduce FastAPI by building a simple application. We shall cover in-depth operations in subsequent chapters.

We'll begin by installing the dependencies required for our application in the todos folder we created earlier. The dependencies are the following:

- fastapi: The framework on which we'll build our application.

- uvicorn: An Asynchronous Server Gateway Interface module to run our application.

First, activate your development environment by running the following command in your project directory:

```
$ source venv/bin/activate
```

Then, install the dependencies as follows:

```
(venv)$ pip install fastapi uvicorn
```

For now, we'll create a new api.py file and create a new instance of FastAPI as follows:

```
from fastapi import FastAPI


app = FastAPI()
```

By instantiating FastAPI in the app variable, we can proceed to create routes. Let's create a welcome route.

A route is created by first defining a decorator to indicate the type of operation, followed by a function containing the operation to be carried out when this route is invoked. In the following example, we'll create a "/" route that only accepts GET requests and returns a welcome message when visited:

```
@app.get("/")
async def welcome() -> dict:
    return { "message": "Hello World"}
```

The next step is to start our application using uvicorn. In your terminal, run the following command:

```
(venv)$ uvicorn api:app --port 8000 --reload
```

In the preceding command, `uvicorn` takes the following arguments:

- `file:instance`: The file containing the instance of FastAPI and the name variable holding the FastAPI instance.

- `--port PORT`: The port the application will be served on.

- `--reload`: An optional argument included to restart the application on every file change.

The command returns the following output:

```
(venv) →  todos uvicorn api:app --port 8080 --reload
INFO:     Will watch for changes in these directories: ['/
Users/youngestdev/Documents/todos']
INFO:     Uvicorn running on http://0.0.0.0:8080 (Press CTRL+C
to quit)
INFO:     Started reloader process [3982] using statreload
INFO:     Started server process [3984]
INFO:     Waiting for application startup.
INFO:     Application startup complete.
```

The next step is to test the application by sending a GET request to the API. In a new terminal, send a GET request using `curl` as follows:

```
$ curl http://0.0.0.0:8080/
```

The response from the application logged in your console will be the following:

```
{"message":"Hello World"}
```

Summary

In this chapter, we have learned how to install the tools required to set up our development environment. We have also built a simple API as an introduction to FastAPI and learned how to create a route in the process.

In the next chapter, you will be introduced to routing in FastAPI. First, you will be introduced to the process of building models to validate request payloads and responses using Pydantic. You will then learn about Path and Query parameters as well as request body, and finally, you will learn how to build a CRUD todo application.

2
Routing in FastAPI

Routing is an essential part of building a web application. Routing in FastAPI is flexible and hassle-free. Routing is the process of handling **HTTP requests** sent from a client to the server. HTTP requests are sent to defined routes, which have defined handlers for processing the requests and responding. These handlers are called route handlers.

By the end of this chapter, you will know how to create routes using the **APIRouter** instance and connect to the main **FastAPI** application. You will also learn what models are and how to use them to validate request bodies. You will also learn what path and query parameters are and how to use them in your FastAPI application. The knowledge of routing in FastAPI is essential in building small- and large-scale applications.

In this chapter, we'll be covering the following topics:

- Routing in FastAPI
- The `APIRouter` class
- Validation using Pydantic models
- Path and query parameters
- Request body
- Building a simple CRUD app

Technical requirements

The code used in this chapter can be found at the `https://github.com/PacktPublishing/Building-Python-Web-APIs-with-FastAPI/tree/main/ch02/todos`.

Understanding routing in FastAPI

A route is defined to accept requests from an HTTP request method and optionally take parameters. When a request is sent to a route, the application checks whether the route is defined before processing the request in the route handler. On the other hand, a route handler is a function that processes the request sent to the server. An example of a route handler is a function that retrieves records from a database when a request is sent to a router via a route.

What are HTTP request methods?

HTTP methods are identifiers for indicating the type of action to be carried out. The standard methods include `GET`, `POST`, `PUT`, `PATCH`, and `DELETE`. You can learn more about HTTP methods at `https://developer.mozilla.org/en-US/docs/Web/HTTP/Methods`.

Routing example

In the *Project scaffolding* section in the previous chapter, we built a single route application. The routing was handled by the `FastAPI()` instance initiated in the app variable:

```
from fastapi import FastAPI


app = FastAPI()


@app.get("/")
async def welcome() -> dict:
    return { "message": "Hello World"}
```

The **uvicorn** tool was pointed to the FastAPI's instance to serve the application:

```
(venv)$ uvicorn api:app --port 8080 --reload
```

Traditionally, the `FastAPI()` instance can be used for routing operations, as seen previously. However, this method is commonly used in applications that require a single path during routing. In a situation where a separate route performing a unique function is created using the `FastAPI()` instance, the application will be unable to run both routes, as uvicorn can only run one entry point.

How then do you handle extensive applications that require a series of routes performing different functions? We'll look at how the **APIRouter** class helps with multiple routing in the next section.

Routing with the APIRouter class

The `APIRouter` class belongs to the FastAPI package and creates path operations for multiple routes. The `APIRouter` class encourages modularity and organization of application routing and logic.

The `APIRouter` class is imported from the `fastapi` package, and an instance is created. The route methods are created and distributed from the instance created, such as the following:

```
from fastapi import APIRouter


router = APIRouter()


@router.get("/hello")
async def say_hello() -> dict:
    return {"message": "Hello!"}
```

Let's create a new path operation with the `APIRouter` class to create and retrieve todos. In the `todos` folder from the previous chapter, create a new file, `todo.py`:

```
(venv)$ touch todo.py
```

We'll start by importing the `APIRouter` class from the fastapi package and creating an instance:

```
from fastapi import APIRouter


todo_router = APIRouter().
```

Next, we'll create a temporary in-app database, alongside two routes for the addition and retrieval of todos:

```
todo_list = []

@todo_router.post("/todo")
async def add_todo(todo: dict) -> dict:
    todo_list.append(todo)
    return {"message": "Todo added successfully"}



@todo_router.get("/todo")
async def retrieve_todos() -> dict:
    return {"todos": todo_list}
```

In the preceding code block, we have created two routes for our todo operations. The first route adds a todo to the todo list via the POST method, and the second route retrieves all the todo items from the todo list via the GET method.

We have completed the path operations for the todo route. The next step is to serve the application to production so that we can test the path operations defined.

The APIRouter class works in the same way as the FastAPI class does. However, uvicorn cannot use the APIRouter instance to serve the application, unlike the FastAPIs. Routes defined using the APIRouter class are added to the fastapi instance to enable their visibility.

To enable the visibility of the todo routes, we'll include the todo_router path operations handler to the primary FastAPI instance using the include_router() method.

> **include_router()**
>
> The include_router(router, …) method is responsible for adding routes defined with the APIRouter class to the main application's instance to enable the routes to become visible.

In api.py, import todo_router from todo.py:

```
from todo import todo_router
```

Include the `todo_router` in the FastAPI application, using the `include_router` method from the **FastAPI** instance:

```python
from fastapi import FastAPI
from todo import todo_router


app = FastAPI()



@app.get("/")
async def welcome() -> dict:
    return {
        "message": "Hello World"
    }


app.include_router(todo_router)
```

With everything in place, start the application from your terminal:

```
(venv)$ uvicorn api:app --port 8000 --reload
```

The preceding command starts our application and gives us a real-time log of our application processes:

```
(venv) →  todos git:(main) X uvicorn api:app --port 8000
--reload
INFO: Will watch for changes in these directories: ['/Users/
youngestdev/Work/Building-Web-APIs-with-FastAPI-and-Python/
ch02/todos']
INFO:     uvicorn running on http://127.0.0.1:8000 (Press
CTRL+C to quit)
INFO:     Started reloader process [4732] using statreload
INFO:     Started server process [4734]
INFO:     Waiting for application startup.
INFO:     Application startup complete.
```

The next step is to test the application by sending a GET request using `curl`:

```
(venv)$ curl http://0.0.0.0:8080/
```

The response from the application logged in your console:

```
{"message":"Hello World"}
```

Next, we check whether the todo routes are functional:

```
(venv)$ curl -X 'GET' \
   'http://127.0.0.1:8000/todo' \
   -H 'accept: application/json'
```

The response from the application logged in your console should be as follows:

```
{
   "todos": []
}
```

The todo route worked! Let's test the POST operation by sending a request to add an item to our todo list:

```
(venv)$ curl -X 'POST' \
   'http://127.0.0.1:8000/todo' \
   -H 'accept: application/json' \
   -H 'Content-Type: application/json' \
   -d '{
   "id": 1,
   "item": "First Todo is to finish this book!"
}'
```

We have the following response:

```
{
   "message": "Todo added successfully."
}
```

We've learned how the APIRouter class works and how to include it in the primary application instance to enable the usage of the path operations defined. The todo routes built in this section lacked models, otherwise known as schemas. In the next section, let's take a look at **Pydantic** models and their use cases.

Validating request bodies using Pydantic models

In FastAPI, request bodies can be validated to ensure only defined data is sent. This is crucial, as it serves to sanitize request data and reduce malicious attacks' risks. This process is known as validation.

A model in FastAPI is a structured class that dictates how data should be received or parsed. Models are created by subclassing Pydantic's `BaseModel` class.

> **What is Pydantic?**
>
> Pydantic is a Python library that handles data validation using Python-type annotations.

Models, when defined, are used as type hints for request body objects and request-response objects. In this chapter, we will only look at using Pydantic models for request bodies.

An example model is as follows:

```
from pydantic import BaseModel


class PacktBook(BaseModel):
    id: int
    Name: str
    Publishers: str
    Isbn: str
```

In the preceding code block above, we defined a `PacktBook` model as a subclass of Pydantic's `BaseModel` class. A variable type hinted to the `PacktBook` class can only take four fields, as defined previously. In the next couple of examples, we see how Pydantic helps in validating inputs.

In our todo application earlier, we defined a route to add an item to our `todo` list. In the route definition, we set the request body to a dictionary:

```
async def add_todo(todo: dict) -> dict:
    ...
```

In the example `POST` request, the data sent was in the following format:

```
{
    "id": id,
    "item": item
}
```

However, an empty dictionary could've also been sent without returning any error. A user can send a request with a body different from the one shown previously. Creating a model with the required request body structure and assigning it as a type to the request body ensures that only the data fields present in the model are passed.

For example, to ensure only the request body contains fields in the preceding example, create a new `model.py` file and add the following code below to it:

```
from Pydantic import BaseModel


class Todo(BaseMode):
    id: int
    item: str
```

In the preceding code block, we have created a Pydantic model that accepts only two fields:

- `id` of type integer
- `item` of type string

Let's go ahead and use the model in our `POST` route. In `api.py`, import the model:

```
from model import Todo
```

Next, replace the request body variable type from `dict` to `Todo`:

```
todo_list = []


@todo_router.post("/todo")
async def add_todo(todo: Todo) -> dict:
    todo_list.append(todo)
    return {"message": "Todo added successfully"}
```

```
@todo_router.get("/todo")
async def retrieve_todos() -> dict:
    return {"todos": todo_list}
```

Let's verify the new request body validator by sending an empty dictionary as the request body:

```
(venv)$ curl -X 'POST' \
  'http://127.0.0.1:8000/todo' \
  -H 'accept: application/json' \
  -H 'Content-Type: application/json' \
  -d '{
}'
```

We get a response, indicating the absence of the id and item field in the request body:

```
{
    "detail": [
      {
        "loc": [
          "body",
          "id"
        ],
        "msg": "field required",
        "type": "value_error.missing"
      },
      {
        "loc": [
          "body",
          "item"
        ],
        "msg": "field required",
        "type": "value_error.missing"
      }
    ]
}
```

Sending a request with correct data returns a successful response:

```
(venv)$ curl -X 'POST' \
  'http://127.0.0.1:8000/todo' \
  -H 'accept: application/json' \
  -H 'Content-Type: application/json' \
  -d '{
  "id": 2,
  "item": "Validation models help with input types"
}'
```

Here is the response:

```
{
  "message": "Todo added successfully."
}
```

Nested models

Pydantic models can also be nested, such as the following:

```
class Item(BaseModel)
    item: str
    status: str


class Todo(BaseModel)
    id: int
    item: Item
```

As a result, a todo of type `Todo` will be represented as the following:

```
{
  "id": 1,
  "item": {
      "item": "Nested models",
      "Status": "completed"
    }
}
```

We have learned what models are, how to create one, and their use cases. We will be using it subsequently in the remaining parts of this book. In the next section, let's look at **path** and **query** parameters.

Path and query parameters

In the previous section, we learned what models are and how they are used to validate request bodies. In this section, you'll learn what path and query parameters are, the role they play in routing, and how to use them.

Path parameters

Path parameters are parameters included in an API route to identify resources. These parameters serve as an identifier and, sometimes, a bridge to enable further operations in a web application.

We currently have routes for adding a todo and retrieving all the todos in our todo application. Let's create a new route for retrieving a single todo by appending the todo's ID as a path parameter.

In todo.py, add the new route:

```
from fastapi import APIRouter, Path
from model import Todo


todo_router = APIRouter()


todo_list = []



@todo_router.post("/todo")
async def add_todo(todo: Todo) -> dict:
    todo_list.append(todo)
    return {
        "message": "Todo added successfully."
    }
```

```
@todo_router.get("/todo")
async def retrieve_todo() -> dict:
    return {
        "todos": todo_list
    }




@todo_router.get("/todo/{todo_id}")
async def get_single_todo(todo_id: int = Path(..., title="The
ID of the todo to retrieve.")) -> dict:
    for todo in todo_list:
        if todo.id == todo_id:
            return {
                "todo": todo
            }
    return {
        "message": "Todo with supplied ID doesn't exist."
    }
```

In the preceding code block, {todo_id} is the path parameter. This parameter enables the application to return a matching todo with the ID passed.

Let's test the route:

```
(venv)$ curl -X 'GET' \
    'http://127.0.0.1:8000/todo/1' \
    -H 'accept: application/json'
```

In the preceding GET request, 1 is the path parameter. Here, we are telling our todo application to return the todo item with 1 ID.

Executing the preceding request results in the following response:

```
{
  "todo": {
    "id": 1,
    "item": "First Todo is to finish this book!"
  }
}
```

FastAPI also provides a Path class that distinguishes path parameters from other arguments present in the route function. The Path class also helps give route parameters more context during the documentation automatically provided by OpenAPI via **Swagger** and **ReDoc** and acts as a validator.

Let's modify the route definition:

```python
from fastAPI import APIRouter, Path

from model import Todo

todo_router = APIRouter()

todo_list = []


@todo_router.post("/todo")
async def add_todo(todo: Todo) -> dict:
    todo_list.append(todo)
    return {
        "message": "Todo added successfully."
    }



@todo_router.get("/todo")
async def retrieve_todo() -> dict:
    return {
        "todos": todo_list
    }



@todo_router.get("/todo/{todo_id}")
async def get_single_todo(todo_id: int = Path(..., title="The
ID of the todo to retrieve")) -> dict:
    for todo in todo_list:
        if todo.id == todo_id:
            return {
                "todo": todo
```

```
        }
    return {
        "message": "Todo with supplied ID doesn't exist."
    }
```

> **Tip – Path(..., kwargs)**
>
> The `Path` class takes a first positional argument set to `None` or ellipsis (...).
> If the first argument is set to an ellipsis (...), the path parameter becomes
> required. The `Path` class also contains arguments used for numerical
> validations if a path parameter is a number. Definitions include `gt` and `le`
> – `gt` means greater than and `le` means less than. When used, the route will
> validate the path parameter against these arguments.

Query parameters

A query parameter is an optional parameter that usually appears after a question mark in
a URL. It is used to filter requests and return specific data based on the queries supplied.

In a route handler function, an argument that isn't homonymous with the path parameter
is a query. You can also define a query by creating an instance of the FastAPI `Query()`
class in the function argument, such as the following:

```
async query_route(query: str = Query(None):
    return query
```

We will be looking at the use cases of the query parameters later on in the book when
we discuss how to build more advanced applications than a todo application.

Now that you have learned how to create routes, validate request bodies, and use path and
query parameters in your FastAPI application, you will learn how these components work
hand in hand to form a request body in the next section.

Request body

In the previous sections, we learned how to use the `APIRouter` class and Pydantic
models for request body validations and discussed path and query parameters.

A request body is data that you send to your API using a routing method such as `POST`
and `UPDATE`.

> **POST and UPDATE**
>
> The POST method is used when an insertion into the server is to be made, and the UPDATE method is used when existing data in the server is to be updated.

Let's take a look at a POST request from earlier on in the chapter:

```
(venv)$ curl -X 'POST' \
  'http://127.0.0.1:8000/todo' \
  -H 'accept: application/json' \
  -H 'Content-Type: application/json' \
  -d '{
  "id": 2,
  "item": "Validation models help with input types"
}'
```

In the preceding request, the request body is as follows:

```
{
  "id": 2,
  "item": "Validation models help with input types.."
}
```

> **Tip**
>
> FastAPI also provides us with a Body() class to provide extra validation.

We have learned about models in FastAPI. They also serve an additional purpose in documenting our API endpoints and request body types. In the next subsection, we will learn about the documentation pages generated by default in FastAPI applications.

FastAPI Automatic Docs

FastAPI generates JSON schema definitions for our models and automatically documents our routes, including their request body type, path and query parameters, and response models. This documentation is of two types:

- **Swagger**
- **ReDoc**

Swagger

The documentation hosted by swagger provides an interactive environment to test our API. You can access it by appending /docs to the application address. In your web browser, visit the http://127.0.0.1:8000/docs URL:

Figure 2.1 – The FastAPI interactive documentation

The interactive documentation allows us to test our methods. Let's add a todo from the interactive documentation:

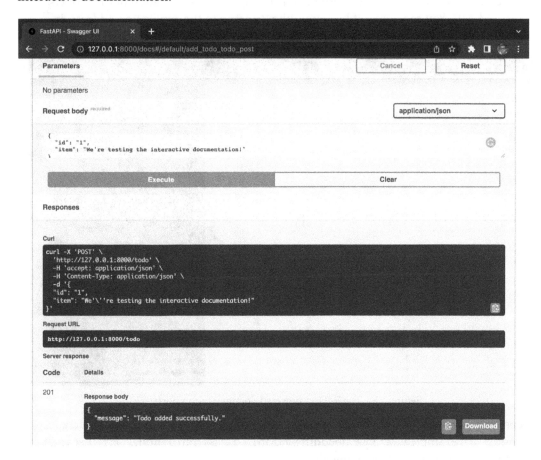

Figure 2.2 – Route test from interactive documentation

Now that we know what the interactive documentation looks like, let's check the documentation generated by ReDoc.

ReDoc

The ReDoc documentation gives a more detailed and direct presentation of the models, routes, and API. You can access it by appending /redoc to the application address. In your web browser, visit the http://127.0.0.1:8000/redoc URL:

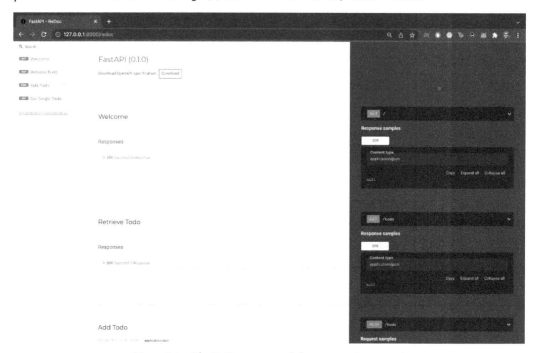

Figure 2.3 – The ReDoc-powered documentation portal

To correctly generate **JSON Schema**, you can set examples of how a user will fill data in the model. An example is set by embedding a Config class into a model class. Let's add an example schema in our Todo model:

```
class Todo(BaseModel):
    id: int
    item: str

    class Config:
        Schema_extra = {
            "Example": {
                "id": 1,
```

```
            "item": "Example schema!"
                                    }

            }
```

Refresh the documentation page for ReDoc and click on **Add Todo** on the left pane. The example is shown in the right pane:

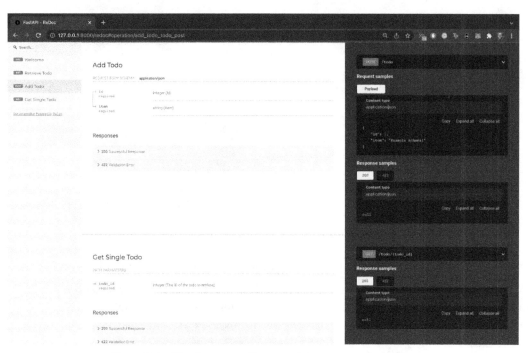

Figure 2.4 – The documentation portal shows example schema

Also in the interactive documentation, the example schema can be seen:

Figure 2.5 – The documentation portal shows example schema

We have learned how to add example schema data to guide users on how to send requests to the API and test the application from Swagger's interactive documentation. The documentation provided by ReDoc isn't left out, as it serves as a knowledge base on how to use the API.

Now that we have learned what the APIRouter class is and how to use it, the request body, path and query parameters, and validating request bodies with Pydantic models, let's update our todo app to include routes for updating and deleting a todo item.

Building a simple CRUD app

We have built routes for **creating** and **retrieving** todos. Let's build the routes for **updating** and **deleting** the added todo. Let's start by creating a model for the request body for the UPDATE route in `model.py`:

```python
class TodoItem(BaseModel):
    item: str

    class Config:
        schema_extra = {
            "example": {
                "item": "Read the next chapter of the book"
            }
        }
```

Next, let's write the route for updating a todo in `todo.py`:

```python
from fastapi import APIRouter, Path
from model import Todo, TodoItem


todo_router = APIRouter()


todo_list = []



@todo_router.post("/todo")
async def add_todo(todo: Todo) -> dict:
    todo_list.append(todo)
    return {
        "message": "Todo added successfully."
    }



@todo_router.get("/todo")
async def retrieve_todo() -> dict:
    return {
        "todos": todo_list
```

```
    }


@todo_router.get("/todo/{todo_id}")
async def get_single_todo(todo_id: int = Path(..., title="The
ID of the todo to retrieve")) -> dict:
    for todo in todo_list:
        if todo.id == todo_id:
            return {
                "todo": todo
            }
    return {
        "message": "Todo with supplied ID doesn't exist."
    }


@todo_router.put("/todo/{todo_id}")
async def update_todo(todo_data: TodoItem, todo_id: int =
Path(..., title="The ID of the todo to be updated")) -> dict:
    for todo in todo_list:
        if todo.id == todo_id:
            todo.item = todo_data.item
            return {
                "message": "Todo updated successfully."
            }
    return {
        "message": "Todo with supplied ID doesn't exist."
    }
```

Let's test the new route. First, let's add a todo:

```
(venv)$ curl -X 'POST' \
  'http://127.0.0.1:8000/todo' \
  -H 'accept: application/json' \
  -H 'Content-Type: application/json' \
  -d '{
  "id": 1,
```

```
    "item": "Example Schema!"
}'
```

Here is the response:

```
(venv)$ {
    "message": "Todo added successfully."
}
```

Next, let's update the todo by sending a PUT request:

```
(venv)$ curl -X 'PUT' \
    'http://127.0.0.1:8000/todo/1' \
    -H 'accept: application/json' \
    -H 'Content-Type: application/json' \
    -d '{
    "item": "Read the next chapter of the book."
}'
```

Here is the response:

```
(venv)$ {
    "message": "Todo updated successfully."
}
```

Let's verify that our todo has indeed been updated:

```
(venv)$ curl -X 'GET' \
    'http://127.0.0.1:8000/todo/1' \
    -H 'accept: application/json'
```

Here is the response:

```
(venv)$ {
    "todo": {
        "id": 1,
        "item": "Read the next chapter of the book"
    }
}
```

From the response returned, we can see that the todo has successfully been updated. Now, let's create the route for deleting a todo and all todos.

In todo.py, update the routes:

```python
@todo_router.delete("/todo/{todo_id}")
async def delete_single_todo(todo_id: int) -> dict:
    for index in range(len(todo_list)):
        todo = todo_list[index]
        if todo.id == todo_id:
            todo_list.pop(index)
            return {
                "message": "Todo deleted successfully."
            }
    return {
        "message": "Todo with supplied ID doesn't exist."
    }



@todo_router.delete("/todo")
async def delete_all_todo() -> dict:
    todo_list.clear()
    return {
        "message": "Todos deleted successfully."
    }
```

Let's test the delete route. First, we add a todo:

```
(venv)$ curl -X 'POST' \
  'http://127.0.0.1:8000/todo' \
  -H 'accept: application/json' \
  -H 'Content-Type: application/json' \
  -d '{
  "id": 1,
  "item": "Example Schema!"
}'
```

Here is the response:

```
(venv)$ {
  "message": "Todo added successfully."
}
```

Next, delete the todo:

```
(venv)$ curl -X 'DELETE' \
  'http://127.0.0.1:8000/todo/1' \
  -H 'accept: application/json'
```

Here is the response:

```
(venv)$ {
  "message": "Todo deleted successfully."
}
```

Let's verify that the todo has been deleted by sending a GET request to retrieve the todo:

```
(venv)$ curl -X 'GET' \
  'http://127.0.0.1:8000/todo/1' \
  -H 'accept: application/json'
```

Here is the response:

```
(venv)$ {
  "message": "Todo with supplied ID doesn't exist.
}
```

In this section, we built a CRUD todo application combining the lessons learned from the preceding sections. By validating our request body, we were able to ensure that proper data is sent to the API. The inclusion of path parameters to our routes also enabled us to retrieve and delete a single todo from our todo list.

Summary

In this chapter, we learned how to use the `APIRouter` class and connect routes defined with it to the primary FastAPI instance. We also learned how to create models for our request bodies and add path and query parameters to our path operations. These models serve as extra validation against improper data types supplied to request body fields. We also built a CRUD todo application to put into practice all that we learned in this chapter.

In the next chapter, you will be introduced to response, response modeling, and error handling in FastAPI. You will first be introduced to the concept of responses and how the knowledge of Pydantic models learned in this chapter helps build models for API responses. You will then learn about status codes and how to use them in your response objects and proper error handling.

3
Response Models and Error Handling

Response models serve as templates for returning data from an API route path. They are built on **Pydantic** to properly render a response from requests sent to the server.

Error handling includes the practices and activities involved in handling errors from an application. These practices include returning adequate error status codes and error messages.

By the end of this chapter, you will know what a response is and what it consists of, and you'll know about error handling and how to handle errors in your FastAPI application. You will also know how to build response models for request responses using Pydantic.

In this chapter, we'll be covering the following topics:

- Responses in FastAPI
- Building a response model
- Error handling

Technical requirements

The code used in this chapter can be found at `https://github.com/PacktPublishing/Building-Python-Web-APIs-with-FastAPI/tree/main/ch03/todos`.

Understanding responses in FastAPI

Responses are an integral part of an API's life cycle. Responses are the feedback received from interacting with an API route via any of the standard HTTP methods. An API response is usually in JSON or XML format, but it can also be in the form of a document. A response consists of a header and a body.

What is a response header?

A response header consists of the request's status and additional information to guide the delivery of the response body. An example of the information contained in the response header is `Content-Type`, which tells the client the content type returned.

What is a response body?

The response body, on the other hand, is the data requested from the server by the client. The response body is determined from the `Content-Type` header variable and the most commonly used one is `application/json`. In the previous chapter, the list of to-dos returned is the response body.

Now that you've learned what responses are and what they consist of, let's take a look at HTTP status codes included in responses in the next section.

Status codes

Status codes are unique short codes issued by a server in response to a client's request. Response status codes are grouped into five categories, each denoting a different response:

- `1XX`: Request has been received.
- `2XX`: The request was successful.
- `3XX`: Request redirected.
- `4XX`: There's an error from the client.
- `5XX`: There's an error from the server.

A complete list of HTTP status codes can be found at https://httpstatuses.com/.

The first digit of a status code defines its category. Common status codes include 200 for a successful request, 404 for request not found, and 500 indicating an internal server error.

The standard practice followed in building web applications, irrespective of the framework, is to return appropriate status codes for individual events. A 400 status code shouldn't be returned for a server error. Likewise, a 200 status code shouldn't be returned for a failed request operation.

Now that you have learned what status codes are, let's learn how to build response models in the next section.

Building response models

We established the purpose of response models at the beginning of this chapter. You also learned how to build models in the previous chapter using Pydantic. Response models are also built on Pydantic but serve a different purpose.

In the definition of route paths, we have the following, for example:

```
@app.get("/todo")
async def retrieve_todo() -> dict:
    return {
        "todos": todo_list
    }
```

The route returns a list of to-dos present in the database. Here's some example output:

```
{
  "todos": [
    {
      "id": 1,
      "item": "Example schema 1!"
    },
    {
      "id": 2,
      "item": "Example schema 2!"
    },
    {
      "id": 3,
```

```
        "item": "Example schema 5!"
    }
  ]
}
```

The route returns all the content stored in the todos array. To specify the information to be returned, we would have to either separate data to be displayed or introduce additional logic. Fortunately, we can create a model containing the fields we want to be returned and add it to our route definition using the `response_model` argument.

Let's update the route that retrieves all the to-dos to return an array of just the to-do items and not the IDs. Let's start by defining a new model class to return a list of to-do items in `model.py`:

```python
from typing import List


class TodoItem(BaseModel):
    item: str

    class Config:
        schema_extra = {
            "example": {
                "item": "Read the next chapter of the book"
            }
        }


class TodoItems(BaseModel):
    todos: List[TodoItem]

    class Config:
        schema_extra = {
            "example": {
                "todos": [
                    {
                        "item": "Example schema 1!"
                    },
                    {
                        "item": "Example schema 2!"
```

```
                }
            ]
        }
    }
```

In the preceding code block, we have defined a new model, TodoItems, which returns a list of variables contained in the TodoItem model. Let's update our route in todo.py by adding a response model to it:

```
from model import Todo, TodoItem, TodoItems

...

@todo_router.get("/todo", response_model=TodoItems)
async def retrieve_todo() -> dict:
    return {
        "todos": todo_list
    }
```

Activate your virtual environment and start your application:

```
$ source venv/bin/activate
(venv)$ uvicorn api:app --host=0.0.0.0 --port 8000 --reload
```

Next, add a new to-do:

```
(venv)$ curl -X 'POST' \
  'http://127.0.0.1:8000/todo' \
  -H 'accept: application/json' \
  -H 'Content-Type: application/json' \
  -d '{
  "id": 1,
  "item": "This todo will be retrieved without exposing my
    ID!"
}'
```

Retrieve the to-dos:

```
(venv)$ curl -X 'GET' \
  'http://127.0.0.1:8000/todo' \
  -H 'accept: application/json'
```

The response received is as follows:

```
{
  "todos": [
    {
      "item": " This todo will be retrieved without
      exposing my ID!"
!"
    }
  ]
}
```

Now that we have learned what response models are and how to use them, we will continue to use them where they fit in subsequent chapters. Let's take a look at error responses and how to handle errors in the next section.

Error handling

Earlier on in this chapter, we learned what status codes are and how they are useful in informing the client about the request status. Requests can return erroneous responses, and these responses can be ugly or have insufficient information about the cause of failure.

Errors from requests can result from attempting to access non-existent resources, protected pages without sufficient permissions, and even server errors. Errors in FastAPI are handled by raising an exception using FastAPI's HTTPException class.

> **What Is an HTTP Exception?**
> An HTTP exception is an event that is used to indicate a fault or issue in the request flow.

The `HTTPException` class takes three arguments:

- `status_code`: The status code to be returned for this disruption
- `detail`: Accompanying message to be sent to the client
- `headers`: An optional parameter for responses requiring headers

In our to-do route path definitions, we return a message when a to-do can't be found. We will be updating it to raise `HTTPException`. `HTTPException` allows us to return an adequate error response code.

In our current application, retrieving a to-do that doesn't exist returns a 200 response status code instead of a 404 response status code on `http://127.0.0.1:8000/docs`:

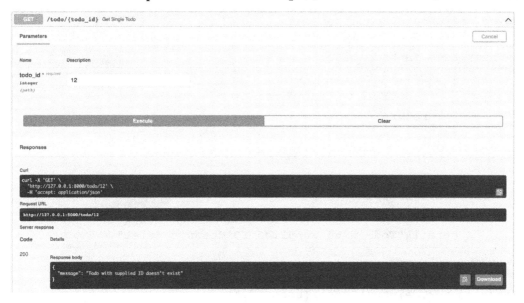

Figure 3.1 – Request returns a 200 response instead of a 404 response

By updating the routes to use the `HTTPException` class, we can return relevant details in our response. In `todo.py`, update the routes for retrieving, updating, and deleting a to-do:

```
from fastapi import APIRouter, Path, HTTPException, status
..
@todo_router.get("/todo/{todo_id}")
async def get_single_todo(todo_id: int = Path(..., title="The
ID of the todo to retrieve.")) -> dict:
    for todo in todo_list:
```

```
            if todo.id == todo_id:
                return {
                    "todo": todo
                }
        raise HTTPException(
            status_code=status.HTTP_404_NOT_FOUND,
            detail="Todo with supplied ID doesn't exist",
        )


@todo_router.put("/todo/{todo_id}")
async def update_todo(todo_data: TodoItem, todo_id: int =
Path(..., title="The ID of the todo to be updated.")) -> dict:
        for todo in todo_list:
            if todo.id == todo_id:
                todo.item = todo_data.item
                return {
                    "message": "Todo updated successfully."
                }

        raise HTTPException(
            status_code=status.HTTP_404_NOT_FOUND,
            detail="Todo with supplied ID doesn't exist",
        )


@todo_router.delete("/todo/{todo_id}")
async def delete_single_todo(todo_id: int) -> dict:
        for index in range(len(todo_list)):
            todo = todo_list[index]
            if todo.id == todo_id:
                todo_list.pop(index)
                return {
                    "message": "Todo deleted successfully."
                }
        raise HTTPException(
            status_code=status.HTTP_404_NOT_FOUND,
```

```
        detail="Todo with supplied ID doesn't exist",
    )
```

Now, let's retry retrieving the non-existent to-do to verify that the right response code is returned:

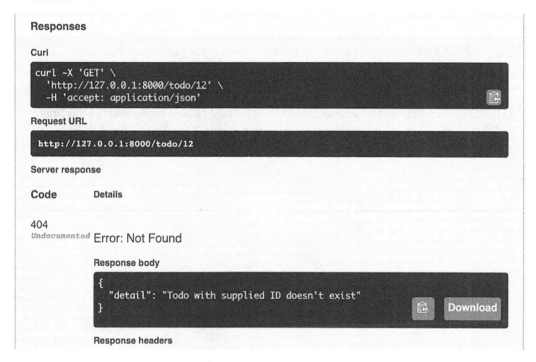

Figure 3.2 – The correct 404 response code displayed

Lastly, we can declare the HTTP status code to override the default status code for successful operations by adding the status_code argument to the decorator function:

```
@todo_router.post("/todo", status_code=201)
async def add_todo(todo: Todo) -> dict:
    todo_list.append(todo)
    return {
        "message": "Todo added successfully."
    }
```

We have learned how to return the right response codes to clients, as well as overriding the default status code, in this section. It is also important to note that the default status code for success is 200.

Summary

In this chapter, we learned what responses and response models are and what is meant by error handling. We also learned about HTTP status codes and why it is important to use them.

We also created response models from the knowledge we gained about creating models from the previous chapter and created a response model to return only the items in the to-do list without their IDs. Lastly, we learned about errors and error handling. We updated our existing routes to return the right response code instead of the default 200 status code.

In the next chapter, you will be introduced to templating FastAPI applications with Jinja. You will first be introduced to the basics needed to get you up and running with Jinja templating, after which you will create a user interface using your templating knowledge for our simple to-do application.

4
Templating in FastAPI

Now that we have learned how to handle responses from requests including errors in the previous chapter, we can proceed to render the responses from the request on a web page. In this chapter, we will learn how to render responses from our API to a web page using templates powered by **Jinja**, which is a templating language written in Python designed to help the rendering process of API responses.

Templating is the process of displaying the data gotten from the API in various formats. Templates act as a frontend component in web applications.

By the end of this chapter, you will be equipped with the knowledge of what templating is and how to use templates to render information from your API. In this chapter, we'll be covering the following topics:

- Understanding Jinja
- Using Jinja2 templates in FastAPI

Technical requirements

The code used in this chapter can be found at `https://github.com/PacktPublishing/Building-Python-Web-APIs-with-FastAPI/tree/main/ch04/todos`.

Understanding Jinja

Jinja is a templating engine written in Python designed to help the rendering process of API responses. In every templating language, there are variables that get replaced with the actual values passed to them when the template is rendered, and there are tags that control the logic of the template.

The Jinja templating engine makes use of curly brackets { } to distinguish its expressions and syntax from regular HTML, text and any other variable in the template file.

The {{ }} syntax is called a **variable block**. The {% %} syntax houses control structures such as **if/else**, **loops**, and **macros**.

The three common syntax blocks used in the Jinja templating language include the following:

- {% ... %} – This syntax is used for statements such as control structures.
- {{ todo.item }} – This syntax is used to print out the values of the expressions passed to it.
- {# This is a great API book! #} – This syntax is used when writing comments and is not displayed on the web page.

Jinja template variables can be of any Python type or object if they can be converted into strings. A model, list, or dictionary type can be passed to the template and have its attributes displayed by placing these attributes in the second block listed previously.

In the next section, we'll be looking at filters. Filters are an important part of every templating engine and in Jinja, filters enable us to execute certain functions such as joining values from a list and retrieving the length of an object, among others.

In the following subsections, we'll be looking at some common features used in Jinja: filters, if statements, loops, macros and template inheritance.

Filters

Despite the similarity between Python and Jinja's syntax, modifications such as joining strings, setting the first character of a string to uppercase, and so on cannot be done using Python's syntax in Jinja. Therefore, to perform such modifications, we have filters in Jinja.

A filter is separated from the variable by a pipe symbol (|) and may entertain optional arguments in parentheses. A filter is defined in this format:

```
{{ variable | filter_name(*args) }}
```

If there are no arguments, the definition becomes the following:

```
{{ variable | filter_name }}
```

Let's take a look at some common filters in the following subsections.

The default filter

The `default` filter variable is used to replace the output of the passed value if it turns out to be `None`:

```
{{ todo.item | default('This is a default todo item') }}
This is a default todo item
```

The escape filter

This filter is used to render raw HTML output:

```
{{ "<title>Todo Application</title>" | escape }}
<title>Todo Application</title>
```

The conversion filters

These filters include `int` and `float` filters used to convert from one data type to another:

```
{{ 3.142 | int }}
3
```

```
{{ 31 | float }}
31.0
```

The join filter

This filter is used to join elements in a list into a string as in Python:

```
{{ ['Packt', 'produces', 'great', 'books!'] | join(' ') }}
Packt produces great books!
```

The length filter

This filter is used to return the length of the object passed. It fulfills the same role as `len()` in Python:

```
Todo count: {{ todos | length }}
Todo count: 4
```

> **Note**
>
> For a full list of filters and to learn more about filters in Jinja, visit
> `https://jinja.palletsprojects.com/en/3.0.x/`
> `templates/#builtin-filters`.

Using if statements

The usage of `if` statements in Jinja is similar to their usage in Python. `if` statements are used in the `{% %}` control blocks. Let's look at an example:

```
{% if todo | length < 5 %}
     You don't have much items on your todo list!
{% else %}
     You have a busy day it seems!
{% endif %}
```

Loops

We can also iterate through variables in Jinja. This could be a list or a general function, such as the following, for example:

```
{% for todo in todos %}
          <b> {{ todo.item }} </b>
{% endfor %}
```

You can access special variables inside a `for` loop, such as `loop.index`, which gives the index of the current iteration. The following is a list of the special variables and their descriptions:

Variable	Description
`loop.index`	The current iteration of the loop (1 indexed)
`loop.index0`	The current iteration of the loop (0 indexed)
`loop.revindex`	The number of iterations from the end of the loop (1 indexed)
`loop.revindex0`	The number of iterations from the end of the loop (0 indexed)
`loop.first`	True if first iteration

Variable	Description
`loop.last`	True if last iteration
`loop.length`	The number of items in the sequence
`loop.cycle`	A helper function to cycle between a list of sequences
`loop.depth`	Indicates how deep in a recursive loop the rendering currently is; starts at level 1
`loop.depth0`	Indicates how deep in a recursive loop the rendering currently is; starts at level 0
`loop.previtem`	The item from the previous iteration of the loop; undefined during the first iteration
`loop.nextitem`	The item from the following iteration of the loop; undefined during the last iteration
`loop.changed(*val)`	True if previously called with a different value (or not called at all)

Macros

A macro in Jinja is a function that return an HTML string. The main use case for macros is to avoid the repetition of code and instead use a single function call. For example, an input macro is defined to reduce the continuous definition of input tags in an HTML form:

```
{% macro input(name, value='', type='text', size=20 %}
    <div class="form">
        <input type="{{ type }}" name="{{ name }}"
        value="{{ value|escape }}" size="{{ size }}">
    </div>
{% endmacro %}
```

Now, to quickly create an input in your form, the macro is called:

```
{{ input('item') }}
```

This will return the following:

```
<div class="form">
    <input type="text" name="item" value="" size="20">
</div>
```

Now that we have learned what macros are, we will proceed to learn what template inheritance is and how it works in FastAPI.

Template inheritance

Jinja's most powerful feature is the inheritance of templates. This feature advances the **don't repeat yourself** (**DRY**) principle and comes in handy in large web applications. Template inheritance is a situation where a base template is defined and child templates can interact, inherit, and replace defined sections of the base template.

> **Note**
>
> You can learn more about Jinja's template inheritance at `https://jinja.palletsprojects.com/en/3.0.x/templates/#template-inheritance`.

Now that you've learned the basics of Jinja's syntax, let's learn how to use templates in FastAPI in the next section.

Using Jinja templates in FastAPI

To get started, we need to install the Jinja package and create a new folder, `templates`, in our project directory. This folder will store all our Jinja files, which are HTML files mixed with Jinja's syntax. Since this book does not focus on user interface design, we will be making use of the CSS `Bootstrap` library and avoid writing our own styles.

The Bootstrap library will be downloaded from the CDN upon page load. However, extra assets can be stored in a different folder. We will look into serving static files in the next chapter.

We'll start by creating the homepage template, which will house the section for creating new todos. The following is a mockup of how we want our homepage template to look:

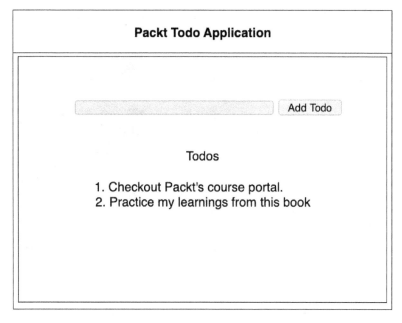

Figure 4.1 – Mockup of our homepage template

1. First, let's install the Jinja package and create the `templates` folder:

```
(venv)$ pip install jinja2
(venv)$ mkdir templates
```

2. In the newly created folder, create two new files, home.html and todo.html:

```
(venv)$ cd templates
(venv)$ touch {home,todo}.html
```

In the preceding command block, we have created two template files:

- home.html for the application's home page
- todo.html for the todo page

In the mockup in *Figure 4.1*, the inner box denotes the todo template while the bigger box is the homepage template.

Before moving on to build our templates, let's configure Jinja in our FastAPI application:

1. Let's modify the POST route of the todo API component, todo.py:

```
from fastapi import APIRouter, Path, HTTPException,
status, Request, Depends
from fastapi.templating import Jinja2Templates
from model import Todo, TodoItem, TodoItems


todo_router = APIRouter()


todo_list = []


templates = Jinja2Templates(directory="templates/")


@todo_router.post("/todo")
async def add_todo(request: Request, todo: Todo =
Depends(Todo.as_form)):
    todo.id = len(todo_list) + 1
    todo_list.append(todo)
    return templates.TemplateResponse("todo.html",
    {
        "request": request,
        "todos": todo_list
    })
```

2. Next, update the GET routes:

```
@todo_router.get("/todo", response_model=TodoItems)
async def retrieve_todo(request: Request):
    return templates.TemplateResponse("todo.html", {
        "request": request,
        "todos": todo_list
    })


@todo_router.get("/todo/{todo_id}")
async def get_single_todo(request: Request, todo_id: int
= Path(..., title="The ID of the todo to retrieve.")):
```

```
    for todo in todo_list:
        if todo.id == todo_id:
            return templates.TemplateResponse(
            "todo.html", {
                "request": request,
                "todo": todo
            })
    raise HTTPException(
        status_code=status.HTTP_404_NOT_FOUND,
        detail="Todo with supplied ID doesn't exist",
    )
```

In the preceding code block, we have configured Jinja to look into the templates directory to serve the templates passed to the templates. TemplateResponse() method.

The POST method for adding a todo has also been updated to include a dependency on the input passed. Dependencies will be discussed in detail in *Chapter 6, Connecting to a Database.*

3. In model.py, add the highlighted code before the Config subclass:

```
from typing import List, Optional

class Todo(BaseModel):
    id: Optional[int]
    item: str

    @classmethod
    def as_form(
        cls,
        item: str = Form(...)
    ):
        return cls(item=item)
```

Now that we have updated our API code, let's write our templates. We'll start by writing the base home.html template in the next step.

4. In `home.html`, we'll start by declaring the document type:

```html
<!DOCTYPE html>
<html lang="en">
    <head>
        <meta charset="UTF-8">
        <meta http-equiv="X-UA-Compatible"
         content="IE=edge">
        <meta name="viewport" content="width=device-
         width, initial-scale=1.0">
        <title>Packt Todo Application</title>
        <link rel="stylesheet" href=
         "https://stackpath.bootstrapcdn.com/
         bootstrap/4.1.0/css/bootstrap.min.css"
         integrity="sha384-9gVQ4dYFwwWSjIDZ
         nLEWnxCjeSWFphJiwGPXr1jddIhOegi
         u1FwO5qRGvFXOdJZ4" crossorigin="anonymous">
        <link rel="stylesheet" href=
         "https://use.fontawesome.com/releases
         /v5.0.10/css/all.css" integrity="sha384-
         +d0P83n9kaQMCwj8F4RJB66tzIwOKmrdb46+porD/
         OvrJ+37WqIM7UoBtwHO6Nlg" crossorigin=
         "anonymous">
    </head>
```

5. The next step is to write the content for the template body. In the template's body, we'll include the name of the application under a `<header></header>` tag, and a link to the child template's `todo_container` wrapped in a block tag. The child template will be written in *step 8*.

    Include the following code just after the `</head>` tag in the `home.html` template file:

```html
<body>
    <header>
        <nav class="navar">
            <div class="container-fluid">
                <center>
```

```
                        <h1>Packt Todo
                        Application</h1>
                    </center>
                </div>
            </nav>
        </header>
        <div class="container-fluid">
            {% block todo_container %}{% endblock %}
        </div>
    </body>
</html>
</html>
```

The highlighted code tells the parent template that the `todo_container` block will be defined by a child template. A child template containing the `todo_container` block and extending the parent template will have its content displayed there.

6. To see the changes, activate your virtual environment and start your application:

```
$ source venv/bin/activate
(venv)$ uvicorn api:app --host=0.0.0.0 --port 8000
--reload
```

7. Open `http://127.0.0.1:8000/todo` to preview the changes:

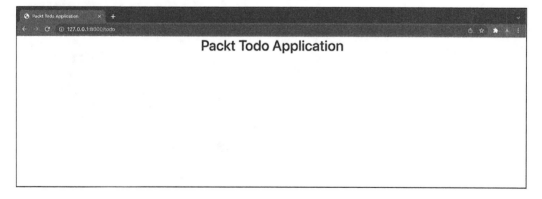

Figure 4.2 – Todo application homepage

8. Next, let's write the todo template in `todo.html`:

```
{% extends "home.html" %}

{% block todo_container %}
<main class="container">
    <hr>
    <section class="container-fluid">
        <form method="post">
            <div class="col-auto">
                <div class="input-group mb-3">
                    <input type="text" name="item"
                     value="{{ item }}" class="form-
                     control" placeholder="Purchase
                     Packt's Python workshop course"
                     aria-label="Add a todo"
                     aria-describedby="button-addon2" />
                    <button class="btn btn-outline-
                     primary" type="submit" id=
                     "button-addon2" data-mdb-ripple-
                     color="dark">
                        Add Todo
                    </button>
                </div>
            </div>
        </form>
    </section>
    {% if todo %}
        <article class="card container-fluid">
            <br/>
            <h4>Todo ID: {{ todo.id }} </h4>
            <p>
                <strong>
                    Item: {{ todo.item  }}
                </strong>
```

```
                </p>
            </article>
    {% else %}
        <section class="container-fluid">
            <h2 align="center">Todos</h2>
            <br>
            <div class="card">
                <ul class="list-group list-group-
                flush">
                    {% for todo in todos %}
                        <li class="list-group-item">
                        {{ loop.index }}. <a href=
                        "/todo/{{loop.index}}"> {{
                        todo.item }} </a>
                        </li>
                    {% endfor %}
                </ul>
            </div>
        {% endif %}
    </section>
</main>
{% endblock %}
```

In the previous code block, the todo template is inheriting the homepage template. We also defined the `todo_container` block, whose content will be displayed in the parent template.

The todo template is used by both routes for retrieving all the todos and for a single todo. As a result, the template renders different content depending on the route used.

In the template, Jinja checks to see whether a todo variable is passed using the `{% if todo %}` block. The todo detail is rendered if a todo variable is passed, otherwise, it renders the content in the `{% else %}` block, which is the list of todos.

9.   Refresh the web browser to view the recent changes:

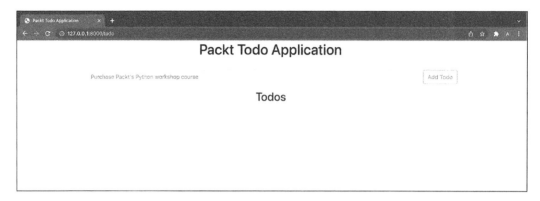

Figure 4.3 – Update todo homepage

10. Let's add a todo to verify that the homepage works as expected:

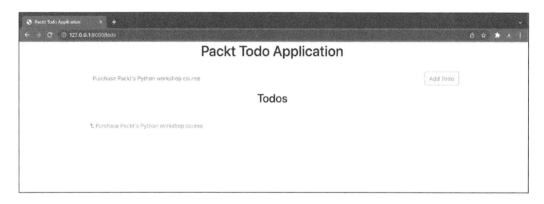

Figure 4.4 – List of todos displayed

11. The todo is clickable. Click on the todo and you should see the following page:

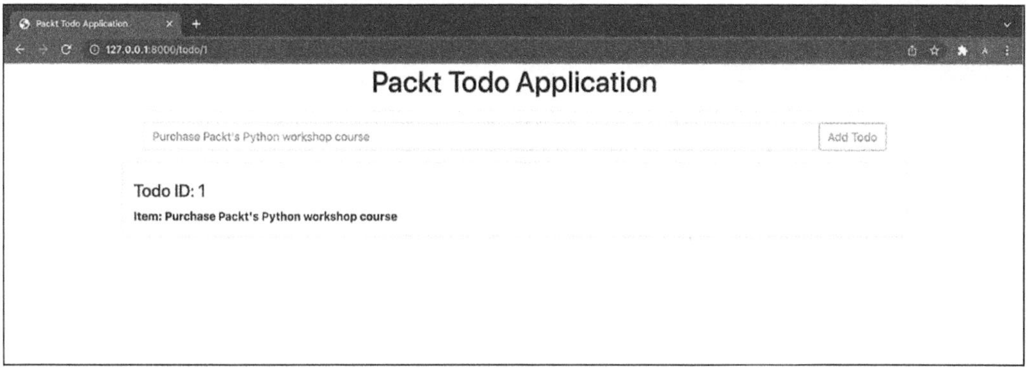

Figure 4.5 – Single todo page

We have successfully added a template to our FastAPI application.

Summary

In this chapter, we learned what templating is, the basics of the Jinja templating system, and how to use it in FastAPI. We made use of the basics learned in the first section of this chapter to decide what content to render. We also learned what template inheritance is and how it works using the homepage and todo templates as examples.

In the next chapter, you will be introduced to structuring applications in FastAPI. In this chapter, you will be building a planner application using the knowledge from this and earlier chapters. You will first be introduced to how applications are structured before proceeding to build the planner application.

Part 2: Building and Securing FastAPI Applications

Upon completing this part, you will be able to build a fully functional and secure application with FastAPI. This part uses the knowledge from the previous part and buttresses it into building a more functional application with higher complexity than the application built in the second chapter. You will also be able to integrate and connect to a SQL and NoSQL (MongoDB) database as well as being able to secure a FastAPI application by the end of this part.

This part comprises the following chapters:

- *Chapter 5, Structuring FastAPI Applications*
- *Chapter 6, Connecting to a Database*
- *Chapter 7, Securing FastAPI Applications*

5
Structuring FastAPI Applications

In the last four chapters, we looked at the basic steps involved in understanding FastAPI and creating a FastAPI application. The application that we have built so far is a single-file todo application that demonstrates the flexibility and power of FastAPI. The key takeaway from the preceding chapters is how easy it is to build an application using FastAPI. However, there is a need for proper structuring of an application with increased complexity and functionalities.

Structuring refers to the arrangement of application components in an organized format, which can be modular to improve the readability of the application's code and content. An application with proper structuring enables faster development, faster debugging, and an overall increase in productivity.

By the end of this chapter, you will be equipped with the knowledge of what structuring is and how to structure your API. In this chapter, you'll be covering the following topics:

- Structuring application routes and models
- Implementing models for a planner API

Technical requirements

The code used in this chapter can be found at `https://github.com/PacktPublishing/Building-Python-Web-APIs-with-FastAPI/tree/main/ch05/planner`.

Structuring in FastAPI applications

For this chapter, we'll be building an event planner. Let's design the application structure to look like this:

```
planner/
    main.py
    database/
        __init__.py
        connection.py
    routes/
        __init__.py
        events.py
        users.py
    models/
        __init__.py
        events.py
        users.py
```

The first step is to create a new folder for the application. It will be named `planner`:

```
$ mkdir planner && cd planner
```

In the newly created `planner` folder, create an entry file, `main.py`, and three subfolders – `database`, `routes`, and `models`:

```
$ touch main.py
$ mkdir database routes models
```

Next, create `__init__.py` in every folder:

```
$ touch {database,routes,models}/__init__.py
```

In the `database` folder, let's create a blank file, `database.py`, which will handle the database abstractions and configurations we'll be using in the next chapter:

```
$ touch database/connection.py
```

In both the `routes` and `models` folders, we'll create two files, `events.py` and `users.py`:

```
$ touch {routes,models}/{events,users}.py
```

Each file has its function, as stated here:

- Files in the `routes` folder:

  - `events.py`: This file will handle routing operations such as creating, updating, and deleting events.

  - `users.py`: This file will handle routing operations such as the registration and signing-in of users.

- Files in the `models` folder:

  - `events.py`: This file will contain the model definition for events operations.

  - `users.py`: This file will contain the model definition for user operations.

Now that we have successfully structured our API and grouped similar files with respect to their functions into components, let's begin the implementation of the application in the next section.

Building an event planner application

In this section, we'll be building an event planner application. In this application, registered users will be able to create, update, and delete events. Events created can be viewed by navigating to the event page created automatically by the application.

Each registered user and event will have a unique ID. This is to prevent conflict in managing users and events with the same ID. In this section, we will not be prioritizing authentication or database management, as this will be discussed in depth in *Chapter 6, Connecting to a Database*, and *Chapter 7, Securing FastAPI Applications*.

To kick-start the development, let us create a virtual environment and activate it in our project directory:

```
$ python3 -m venv venv
$ source venv/bin/activate
```

Next, let's install the application dependencies:

```
(venv)$ pip install fastapi uvicorn "pydantic[email]"
```

Lastly, save the requirements into requirements.txt:

```
(venv)$ pip freeze > requirements.txt
```

Now that we have successfully installed our dependencies and set up our development environment, let's implement the application's models next.

Implementing the models

Let's look at the steps for implementing our model:

1. The first step in building our application is to define the models for the event and user. The models describe how data will be stored, inputted, and represented in our application. The following diagram shows the modeling for both the user and event as well as their relationship:

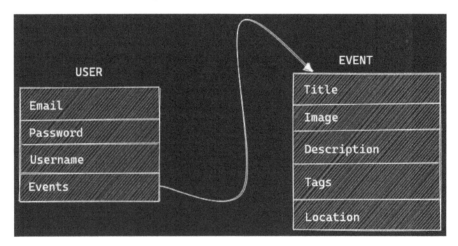

Figure 5.1 – The model flow for the user and event

As shown in the previous model diagram, each user will have an **Events** field, which is a list of the events they have ownership of.

2. Let's define the `Event` model in `models/events.py`:

```python
from pydantic import BaseModel
from typing import List


class Event(BaseModel):
    id: int
    title: str
    image: str
    description: str
    tags: List[str]
    location: str
```

3. Let's define a `Config` subclass under the `Event` class to show an example of what the model data will look like when we visit the documentation:

```python
class Config:
    schema_extra = {
        "example": {
            "title": "FastAPI Book Launch",
            "image": "https:
            //linktomyimage.com/image.png",
            "description": "We will be discussing
            the contents of the FastAPI book in
            this event. Ensure to come with your
            own copy to win gifts!",
            "tags": ["python", "fastapi", "book",
            "launch"]
            "location": "Google Meet"
        }
    }
```

Our event model in the first block of code contains five fields:

- The event title

- A link to the event image banner

- The description of the event

- Event tags for grouping

- The location of the event

In the second block of code, we define example event data. This is aimed at guiding us when creating a new event from our API.

4. Now that we have our event model defined, let's define the User model:

```
from pydantic import BaseModel, EmailStr
from typing import Optional, List
from models.events import Event


class User(BaseModel):
    email: EmailStr
    password: str
    events: Optional[List[Event]]
```

Our User model defined previously contains the following fields:

- The user's email

- The user's password

- A list of events created by the user, which is empty by default

5. Now that we have defined our User model, let's create an example that indicates how the user data is stored and set:

```
class Config:
    schema_extra = {
        "example": {
            "email": fastapi@packt.com,
            "username": "strong!!!",
            "events": [],
        }
    }
```

6. Next, we'll create a new model, NewUser, which inherits from the User model; this new model will be used as the data type when registering a new user. The User model will be used as response models where we do not want to interact with the password, reducing the amount of work to be done.

7. Lastly, let's implement a model for signing users in:

```python
class UserSignIn(BaseModel):
    email: EmailStr
    password: str

    class Config:
        schema_extra = {
            "example": {
                "email": fastapi@packt.com,
                "password": "strong!!!",
                "events": [],
            }
        }
```

Now that we have successfully implemented our models, let's implement the routes in the next section.

Implementing routes

The next step in building our application is to set up the routing system of our API. We'll be designing the routing system for events and users. The user route will consist of sign-in, sign-out, and sign-up routes. An authenticated user will have access to the routes for creating, updating, and deleting an event, while the public can view the event once it has been created. The following diagram shows the relationship between both routes:

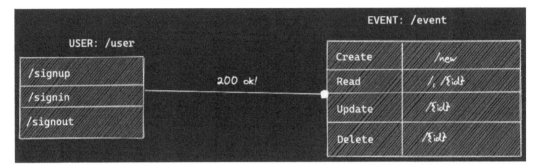

Figure 5.2 – The routes for user and event operations

Let's look at both routes in detail next.

User routes

Now that we have a clear idea of what routes to implement from *Figure 5.2*, we'll start by defining the user routes in users.py. Let's look at the steps:

1. Start by defining a basic sign-up route:

    ```
    from fastapi import APIRouter, HTTPException, status
    from models.user import User, UserSignIn

    user_router = APIRouter(
        tags=["User"]
    )

    users = {}

    @user_router.post("/signup")
    async def sign_new_user(data: NewUser) -> dict:
        if data.email in users:
            raise HTTPException(
    ```

```
            status_code=status.HTTP_409_CONFLICT,
            detail="User with supplied username
            exists"
        )
    users[data.email] = data
    return {
        "message": "User successfully registered!"
    }
```

In the sign-up route defined previously, we're making use of an in-app database. (We'll be introducing a database in *Chapter 6, Connecting to a Database*.)

The route checks whether a user with a similar email address exists in the database before adding a new one.

2. Let's implement the sign-in route:

```
@user_router.post("/signin")
async def sign_user_in(user: UserSignIn) -> dict:
    if users[user.email] not in users:
        raise HTTPException(
            status_code=status.HTTP_404_NOT_FOUND
            detail="User does not exist"
        )
    if users[user.email].password != user.password:
            raise HTTPException(
            status_code=status.HTTP_403_FORBIDDEN
            detail="Wrong credentials passed"
        )


    return {
        "message": "User signed in successfully"
    }
```

In this route, the first step taken is to check whether such a user exists in the database, and if the user doesn't exist, an exception is raised. If the user exists, the application proceeds to check whether the passwords match before returning a successful message or an exception.

In our routes, we're storing the passwords plainly without any encryption. This is used only for demonstration purposes and is a wrong practice in software engineering in general. Proper storage mechanisms such as encryption will be discussed in *Chapter 6, Connecting to a Database*, where our application will move from an in-app database to a real database.

3. Now that we have defined the routes for the user operations, let's register them in `main.py` and start our application. Let's start by importing our libraries and the user routes definition:

```
from fastapi import FastAPI
from routes.user import user_router


import uvicorn
```

4. Next, let's create a `FastAPI` instance and register the route and the application:

```
app = FastAPI()

# Register routes

app.include_router(user_router, prefix="/user")

if __name__ == "__main__":
    uvicorn.run("main:app", host="0.0.0.0", port=8080,
    reload=True)
```

In this block of code, we have created an instance of `FastAPI` and registered the route.

5. Next, we make use of the `uvicorn.run()` method to start our application on port `8080` and set the hot reload to `True`. In the terminal, start the application:

```
(venv)$ python main.py
INFO:      Will watch for changes in these directories:
['/Users/youngestdev/Work/Building-Web-APIs-with-FastAPI-
and-Python/ch05/planner']
INFO:      Uvicorn running on http://0.0.0.0:8080 (Press
CTRL+C to quit)
INFO:      Started reloader process [6547] using
statreload
```

```
INFO:       Started server process [6550]
INFO:       Waiting for application startup.
INFO:       Application startup complete.
```

6. Now that our application has started successfully, let's test the user routes we implemented. We'll start by signing up a user:

```
(venv)$ curl -X 'POST' \
    'http://0.0.0.0:8080/user/signup' \
    -H 'accept: application/json' \
    -H 'Content-Type: application/json' \
    -d '{
    "email": "fastapi@packt.com",
    "password": "Stro0ng!",
    "username": "FastPackt"
}'
```

The preceding request returns a response:

```
{
    "message": "User successfully registered!"
}
```

7. The preceding response indicates the success of the operation performed. Let's test the sign-in route:

```
(venv)$ curl -X 'POST' \
    'http://0.0.0.0:8080/user/signin' \
    -H 'accept: application/json' \
    -H 'Content-Type: application/json' \
    -d '{
    "email": "fastapi@packt.com",
    "password": "Stro0ng!"
}'
```

The preceding response to the request is as follows:

```
{
    "message": "User signed in successfully"
}
```

8. If we pass a wrong password, our application should return a different message:

```
(venv)$ curl -X 'POST' \
  'http://0.0.0.0:8080/user/signin' \
  -H 'accept: application/json' \
  -H 'Content-Type: application/json' \
  -d '{
  "email": "fastapi@packt.com",
  "password": "password!"
}'
```

The response from the preceding request is as follows:

```
{
    "detail": "Wrong credential passed"
}
```

We can also view our routes from the interactive documentation provided by FastAPI, which is powered by Swagger. Let's visit http://0.0.0.0:8080/docs in our browser to gain access to the interactive documentation:

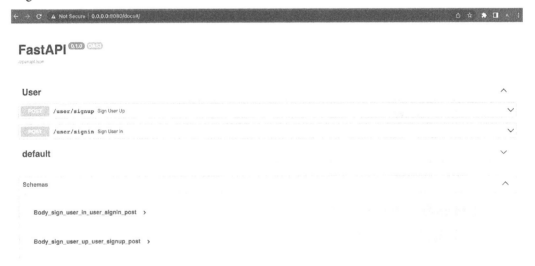

Figure 5.3 – The planner application viewed on the interactive documentation page

Now that we have successfully implemented the user routes, let's implement the routes for event operations in the next section.

Events routes

With the user routes in place, the next step is to implement the routes for event operations. Let's look at the steps:

1. Start by importing the dependencies and defining the event router:

    ```
    from fastapi import APIRouter, Body, HTTPException,
    status
    from models.events import Event
    from typing import List


    event_router = APIRouter(
        tags=["Events"]
    )


    events = []
    ```

2. The next step is to define the route to retrieve all the events and an event matching a supplied ID in the database:

    ```
    @event_router.get("/", response_model=List[Event])
    async def retrieve_all_events() -> List[Event]:
        return events


    @event_router.get("/{id}", response_model=Event)
    async def retrieve_event(id: int) -> Event:
        for event in events:
            if event.id == id:
                return event
        raise HTTPException(
            status_code=status. HTTP_404_NOT_FOUND,
            detail="Event with supplied ID does not exist"
        )
    ```

In the second route, we're raising an HTTP_404_NOT_FOUND exception when an event with the supplied ID doesn't exist.

3. Let's implement the routes to create an event, delete a single event and delete all events contained in the database:

```
@event_router.post("/new")
async def create_event(body: Event = Body(...)) -> dict:
    events.append(body)
    return {
        "message": "Event created successfully"
    }


@event_router.delete("/{id}")
async def delete_event(id: int) -> dict:
    for event in events:
        if event.id == id:
            events.remove(event)
            return { "message": "Event deleted
            successfully" }
    raise HTTPException(
        status_code=status. HTTP_404_NOT_FOUND,
        detail="Event with supplied ID does not exist"
    )
@event_router.delete("/")
async def delete_all_events() -> dict:
    events.clear()
    return {
        "message": "Events deleted successfully"
    }
```

We have successfully implemented the routes for events. The UPDATE route will be implemented in *Chapter 6, Connecting to a Database*, where we'll port our application to use an actual database.

4. Now that we have implemented the routes, let's update our route configuration to include the event route in `main.py`:

```
from fastapi import FastAPI
from routes.user import user_router
from routes.events import event_router
```

```
import uvicorn
app = FastAPI()

# Register routes

app.include_router(user_router, prefix="/user")
app.include_router(event_router, prefix="/event")

if __name__ == "__main__":
    uvicorn.run("main:app", host="0.0.0.0", port=8080,
    reload=True)
```

The application automatically reloads on every change. Let's test the routes:

- The GET route – the following operation returns an empty array, telling us that no data is present:

```
(venv)$ curl -X 'GET' \
  'http://0.0.0.0:8080/event/' \
  -H 'accept: application/json'
[]
```

Let's add data to our array next.

- The POST route – in the terminal, execute the following command:

```
(venv)$ curl -X 'POST' \
  'http://0.0.0.0:8080/event/new' \
  -H 'accept: application/json' \
  -H 'Content-Type: application/json' \
  -d '{
"id": 1,
"title": "FastAPI Book Launch",
"image": "https://linktomyimage.com/image.png",
"description": "We will be discussing the contents
of the FastAPI book in this event.Ensure to come
with your own copy to win gifts!",
"tags": [
  "python",
```

```
        "fastapi",
        "book",
        "launch"
    ],
    "location": "Google Meet"
}'
```

Here is the response:

```
{
    "message": "Event created successfully"
}
```

This operation was successful from the response received. Now, let's try to retrieve the specific event we just created:

- The GET route:

```
(venv)$ curl -X 'GET' \
    'http://0.0.0.0:8080/event/1' \
    -H 'accept: application/json'
```

Here is the response:

```
{
    "id": 1,
    "title": "FastAPI BookLaunch",
    "image": "https://linktomyimage.com/image.png",
    "description": "We will be discussing the contents
    of the FastAPI book in this event.Ensure to come
    with your own copy to win gifts!",
    "tags": [
        "python",
        "fastapi",
        "book",
        "launch"
    ],
    "location": "Google Meet"
}
```

Lastly, let's delete the event to confirm that the event route is working:

- The DELETE route – in the terminal, execute the following command:

```
(venv)$ curl -X 'DELETE' \
  'http://0.0.0.0:8080/event/1' \
  -H 'accept: application/json'
```

Here is the response:

```
{
    "message": "Event deleted successfully"
}
```

If I retry the same command, I get the following response:

```
(venv)$ {
    "detail": "Event with supplied ID does not exist"
}
```

We have successfully implemented the routes and models for our planner application. We have also tested them to assess their working status.

Summary

In this chapter, we learned how to structure a FastAPI application, and implement routes, and models for an event-planning application. We made use of the basics of routing and the knowledge of routing and modeling that we learned in an earlier chapter.

In the next chapter, you will be introduced to connecting your application to SQL and NoSQL databases. You will continue building the event planner application by improving the existing application and adding more features. Before that, you will be introduced to what databases are, the different types, and how to use both (SQL and NoSQL) in a FastAPI application.

6
Connecting to a Database

In the last chapter, we looked at how to structure a FastAPI application. We successfully implemented some routes and models for our application and tested the endpoints. However, the application still uses an in-app database to store the events. In this chapter, we will migrate the application to use a proper database.

A database can be simply referred to as a storehouse for data. In this context, a database enables us to store data permanently, as opposed to an in-app database, which is wiped off upon any app restart or crash. **A database is a table housing columns, referred to as fields, and rows, referred to as records.**

By the end of this chapter, you will be equipped with the knowledge of how to connect a FastAPI application to a database. This chapter will explain how to connect to a SQL database using **SQLModel** and a **MongoDB** database via **Beanie**. (However, the application will make use of MongoDB as its primary database in later chapters.) In this chapter, you'll be covering the following topics:

- Setting up SQLModel
- CRUD operations on a SQL database using SQLModel
- Setting up MongoDB
- CRUD operations on MongoDB using Beanie

Technical requirements

To follow along, the MongoDB database component is required. The installation procedures for your operating system can be found in their official documentation. The code used in this chapter can be found at `https://github.com/PacktPublishing/Building-Python-Web-APIs-with-FastAPI/tree/main/ch06/planner`.

Setting up SQLModel

The first step to integrate a SQL database into our planner application is to install the SQLModel library. The SQLModel library was built by the creator of FastAPI and is powered by Pydantic and SQLAlchemy. Support from Pydantic will make it easy for us to define models, as we learned in *Chapter 3*, *Response Models and Error Handling*.

Since we'll be implementing both SQL and NoSQL databases, we'll create a new GitHub branch for this section. In your terminal, navigate to the project directory, initialize a GitHub repository, and commit the existing files:

```
$ git init
$ git add database models routes main.py
$ git commit -m "Committing bare application without a
database"
```

Next, create a new branch:

```
$ git checkout -b planner-sql
```

Now, we are ready to set up SQLModel in our application. From your terminal, activate the virtual environment and install the SQLModel library:

```
$ source venv/bin/activate
(venv)$ pip install sqlmodel
```

Before diving into adding a database to our planner application, let's look at some of the methods contained in SQLModel that we'll be using in this chapter.

Tables

A table is essentially an object that contains data stored in a database – for example, events data will be stored in an event table. The table will consist of columns and rows where the data will eventually be stored.

To create a table using SQLModel, a table model class is first defined. As with Pydantic models, the table is defined but, this time around, as subclasses of the SQLModel class. The class definition also takes another config variable, `table`, to indicate that this class is a SQLModel table.

The variables defined in the class will represent the columns by default unless denoted as a field. Let's look at how the event table will be defined:

```
class Event(SQLModel, table=True):
    id: Optional[int] = Field(default=None,
    primary_key=True)
    title: str
    image: str
    description: str
    location: str
    tags: List[str]
```

In this `table` class, all the variables defined are columns except `id`, which has been defined as a field. Fields are denoted using the `Field` object from the SQLModel library. The `id` field is also the primary key in the database table.

> **What Is a Primary Key?**
> A primary key is a unique identifier for a record contained in a database table.

Now that we have learned what tables are and how to create them, let's look at rows in the next section.

Rows

Data sent to a database table is stored in rows under specified columns. To insert data into the rows and store them, an instance of the table is created and the variables are filled with the desired input. For example, to insert event data into the events table, we'll create an instance of the model first:

```
new_event = Event(title="Book Launch",
                image="src/fastapi.png",
                description="The book launch event will
                be held at Packt HQ, Packt city",
```

```
                    location="Google Meet",
                    tags=["packt", "book"])
```

Next, we create a database transaction using the `Session` class:

```
with Session(engine) as session:
    session.add(new_event)
    session.commit()
```

The preceding operation may seem alien to you. Let's look at what the **Session** class is and what it does.

Sessions

A session object handles the interaction from code to a database. It primarily acts as an intermediary in executing operations. The `Session` class takes an argument that is the instance of a SQL engine.

Now that we have learned how tables and rows are created, we will look at how a database is created. Some of the methods of the `session` class we'll be using in this chapter include the following:

- `add()`: This method is responsible for adding a database object to memory pending further operations. In the previous code block, the `new_event` object is added to the session's memory, waiting to be committed into the database by the `commit()` method.

- `commit()`: This method is responsible for flushing transactions present in the session.

- `get()`: This method takes two parameters – the model and the ID of the document requested. This method is used to retrieve a single row from a database.

Now that we know how to create tables, rows, and columns, as well as insert data using the `Session` class, let's move on to creating a database and performing CRUD operations in the next section.

Creating a database

In SQLModel, connecting to a database is done via a SQLAlchemy engine. The engine is created by the `create_engine()` method, imported from the SQLModel library.

The create_engine() method takes the database URL as the argument. The database URL is in the form of sqlite:///database.db or sqlite:///database.sqlite. It also takes an optional argument, echo, which when set to True prints out the SQL commands carried out when an operation is executed.

However, the create_engine() method alone isn't sufficient to create a database file. To create the database file, the SQLModel.metadata.create_all(engine) method whose argument is an instance of the create_engine() method is invoked, such as the following:

```
database_file = "database.db"
engine = create_engine(database_file, echo=True)
SQLModel.metadata.create_all(engine)
```

The create_all() method creates the database as well as the tables defined. It is important to note that the file containing the tables is imported into the file where the database connection takes place.

In our planner application, we perform CRUD operations for events. In the database folder, create the following file:

connection.py

In this file, we'll configure the necessary data for the database:

```
(venv) $ touch database/connection.py
```

Now that we have created the database connection file, let's create the functions required to connect our application to the database:

1. We'll start by updating the events model class defined in models/events.py to a SQLModel table model class:

    ```
    from sqlmodel import JSON, SQLModel, Field, Column
    from typing import Optional, List


    class Event(SQLModel, table=True):
        id: int = Field(default=None, primary_key=True)
        title: str
        image: str
        description: str
        tags: List[str] = Field(sa_column=Column(JSON))
    ```

```
    location: str

    class Config:
        arbitrary_types_allowed = True
        schema_extra = {
            "example": {
                "title": "FastAPI Book Launch",
                "image": "https:
                //linktomyimage.com/image.png",
                "description": "We will be discussing
                the contents of the FastAPI book in
                this event. Ensure to come with your
                own copy to win gifts!",
                "tags": ["python", "fastapi", "book",
                "launch"],
                "location": "Google Meet"
            }
        }
```

In this code block, we have modified the original model class to become a SQL table class.

2. Let's add another SQLModel class that'll be used as the body type during UPDATE operations:

```
class EventUpdate(SQLModel):
    title: Optional[str]
    image: Optional[str]
    description: Optional[str]
    tags: Optional[List[str]]
    location: Optional[str]

    class Config:
        schema_extra = {
            "example": {
                "title": "FastAPI Book Launch",
                "image": "https:
```

```
//linktomyimage.com/image.png",
    "description": "We will be discussing
    the contents of the FastAPI book in
    this event. Ensure to come with your
    own copy to win gifts!",
    "tags": ["python", "fastapi", "book",
    "launch"],
    "location": "Google Meet"
    }
}
```

3. Next, let's define the configuration needed to create our database and table in
    `connection.py`:

```python
from sqlmodel import SQLModel, Session, create_engine
from models.events import Event

database_file = "planner.db"
database_connection_string = f"sqlite:///{database_file}"
connect_args = {"check_same_thread": False}
engine_url = create_engine(database_connection_string,
echo=True, connect_args=connect_args)

def conn():
    SQLModel.metadata.create_all(engine_url)

def get_session():
    with Session(engine_url) as session:
        yield session
```

In this code block, we start by defining the dependencies as well as importing the
table model class. Next, we create the variable holding the location of the database
file (which will be created if it doesn't exist), the connection string, and an instance
of the SQL database created. In the `conn()` function, we instruct SQLModel to
create the database as well as the table present in the file, `Events`, and to persist the
session in our application, `get_session()` is defined.

4. Next, let's instruct our application to create a database when it is started. Update `main.py` with the following code:

```python
from fastapi import FastAPI
from fastapi.responses import RedirectResponse
from database.connection import conn

from routes.users import user_router
from routes.events import event_router

import uvicorn

app = FastAPI()

# Register routes

app.include_router(user_router,  prefix="/user")
app.include_router(event_router, prefix="/event")

@app.on_event("startup")
def on_startup():
    conn()

@app.get("/")
async def home():
    return RedirectResponse(url="/event/")

if __name__ == '__main__':
    uvicorn.run("main:app", host="0.0.0.0", port=8080,
    reload=True)
```

The database will be created once the application starts. In the startup event, we have called the `conn()` function responsible for creating the database. Start the application in your terminal and you should see the output in your console, indicating that the database has been created as well as the table:

```
●  ●  ●                        python main.py                              ⌥⌘1
INFO:     Will watch for changes in these directories: ['/Users/youngestdev/Work/Building-Web-APIs-with-FastAPI-a
nd-Python/ch06/planner']
INFO:     Uvicorn running on http://0.0.0.0:8080 (Press CTRL+C to quit)
INFO:     Started reloader process [5668] using statreload
INFO:     Started server process [5671]
INFO:     Waiting for application startup.
2022-04-06 16:25:36,230 INFO sqlalchemy.engine.Engine BEGIN (implicit)
2022-04-06 16:25:36,230 INFO sqlalchemy.engine.Engine PRAGMA main.table_info("event")
2022-04-06 16:25:36,230 INFO sqlalchemy.engine.Engine [raw sql] ()
2022-04-06 16:25:36,230 INFO sqlalchemy.engine.Engine PRAGMA temp.table_info("event")
2022-04-06 16:25:36,231 INFO sqlalchemy.engine.Engine [raw sql] ()
2022-04-06 16:25:36,231 INFO sqlalchemy.engine.Engine
CREATE TABLE event (
        tags JSON,
        id INTEGER,
        title VARCHAR NOT NULL,
        image VARCHAR NOT NULL,
        description VARCHAR NOT NULL,
        location VARCHAR NOT NULL,
        PRIMARY KEY (id)
)


2022-04-06 16:25:36,231 INFO sqlalchemy.engine.Engine [no key 0.00003s] ()
2022-04-06 16:25:36,232 INFO sqlalchemy.engine.Engine COMMIT
```

Figure 6.1 – The planner database and event table created successfully

The SQL commands displayed in the terminal are there because of setting `echo` to `True` when creating the database engine. Now that we have successfully created the database, let's update our events' CRUD operation routes to use the database.

Creating events

Let's look at the steps:

1. In `routes/events.py`, update the imports to include the Event table model class as well as the `get_session()` function. The `get_session()` function is imported so that the routes can access the session object created:

    ```
    from fastapi import APIRouter, Depends, HTTPException,
    Request, status
    from database.connection import get_session
    from models.events import Event, EventUpdate
    ```

> **What Is Depends?**
>
> The `Depends` class is responsible for exercising dependency injection in FastAPI applications. The `Depends` class takes a truth source such as a function as an argument and is passed as a function argument in a route, mandating that the dependency condition be satisfied before any operation can be executed.

2. Next, let's update the `POST` route function responsible for creating a new event, `create_event()`:

```
@event_router.post("/new")
async def create_event(new_event: Event,
session=Depends(get_session)) -> dict:
    session.add(new_event)
    session.commit()
    session.refresh(new_event)

    return {
        "message": "Event created successfully"
    }
```

In this code block, we have indicated that the session object required to execute database transactions is dependent on the `get_session()` function we created earlier.

In the function body, the data is added to the session and then committed to the database, after which the database is refreshed.

3. Let's test the routes to preview changes:

```
(venv)$ curl -X 'POST' \
  'http://0.0.0.0:8080/event/new' \
  -H 'accept: application/json' \
  -H 'Content-Type: application/json' \
  -d '{
  "title": "FastAPI Book Launch",
  "image": "fastapi-book.jpeg",
  "description": "We will be discussing the contents
  of the FastAPI book in this event. Ensure to come
  with your own copy to win gifts!",
```

```
    "tags": [
      "python",
      "fastapi",
      "book",
      "launch"
    ],
    "location": "Google Meet"
  }'
```

A successful response is returned:

```
{
    "message": "Event created successfully"
}
```

If the operation failed to execute, an exception will be thrown by the library.

Read events

Let's update the GET route that retrieves the list of events to pull data from the database:

```
@event_router.get("/", response_model=List[Event])
async def retrieve_all_events(session=Depends(get_session)) ->
List[Event]:
    statement = select(Event)
    events = session.exec(statement).all()
    return events
```

Likewise, the route to display an event's data when retrieved by its ID is also updated:

```
@event_router.get("/{id}", response_model=Event)
async def retrieve_event(id: int, session=Depends(get_session))
-> Event:
    event = session.get(Event, id)
    if event:
        return event
    raise HTTPException(
        status_code=status.HTTP_404_NOT_FOUND,
        detail="Event with supplied ID does not exist"
```

```
    )

    raise HTTPException(
        status_code=status.HTTP_404_NOT_FOUND,
        detail="Event with supplied ID does not exist"
    )
```

The response model for both routes has been set to the model class. Let's test both routes by first sending a GET request to retrieve the list of the events:

```
(venv)$ curl -X 'GET' \
  'http://0.0.0.0:8080/event/' \
  -H 'accept: application/json'
```

We get a response:

```
[
  {
    "id": 1,
    "title": "FastAPI Book  Launch",
    "image": "fastapi-book.jpeg",
    "description": "We will be discussing the contents of
    the FastAPI book in this event.Ensure to come with your
    own copy to win gifts!",
    "tags": [
      "python",
      "fastapi",
      "book",
      "launch"
    ],
    "location": "Google Meet"
  }
}
```

Next, let's retrieve the event by its ID:

```
(venv)$ curl -X 'GET' \
  'http://0.0.0.0:8080/event/1' \
  -H 'accept: application/json'
```

```
}
{
  "id": 1,
  "title": "FastAPI Book Launch",
  "image": "fastapi-book.jpeg",
  "description": "The launch of the FastAPI book will hold
  on xyz.",
  "tags": [
    "python",
    " fastapi"
  ],
  "location": "virtual"
}
```

With the READ operations successfully implemented, let's add an edit feature for our application.

Update events

Let's add the UPDATE route in `routes/events.py`:

```
@event_router.put("/edit/{id}", response_model=Event)
async def update_event(id: int, new_data: EventUpdate,
session=Depends(get_session)) -> Event:
```

In the function body, add the following block of code to retrieve the existing event and handle event changes:

```
    event = session.get(Event, id)
    if event:
        event_data = new_data.dict(exclude_unset=True)
        for key, value in event_data.items():
            setattr(event, key, value)
        session.add(event)
        session.commit()
        session.refresh(event)

        return event
    raise HTTPException(
```

```
              status_code=status.HTTP_404_NOT_FOUND,
              detail="Event with supplied ID does not exist"
    )
```

In the preceding code block, we check whether an event is present before proceeding to update the event data. Once the event has been updated, the updated data is returned. Let's update the existing article's title:

```
(venv)$ curl -X 'PUT' \
  'http://0.0.0.0:8080/event/edit/1' \
  -H 'accept: application/json' \
  -H 'Content-Type: application/json' \
  -d '{
  "title": "Packt'\''s FastAPI book launch II"
}'
```

```
{
  "id": 1,
  "title": "Packt's FastAPI book launch II",
  "image": "fastapi-book.jpeg",
  "description": "The launch of the FastAPI book will hold
on xyz.",
  "tags": ["python", "fastapi"],
  "location": "virtual" }
```

Now that we have added the update functionality, let's quickly add a delete operation in the next section.

Delete event

In events.py, update the delete route defined earlier:

```
@event_router.delete("/delete/{id}")
async def delete_event(id: int, session=Depends(get_session))
-> dict:
    event = session.get(Events, id)
    if event:
        session.delete(event)
```

```
    session.commit()

    return {
        "message": "Event deleted successfully"
    }

raise HTTPException(
    status_code=status.HTTP_404_NOT_FOUND,
    detail="Event with supplied ID does not exist"
)
```

In this code block, the function checks whether an event whose ID has been supplied exists and then deletes it from the database. Once the operation has been executed, a successful message is returned and an exception thrown if the event doesn't exist. Let's delete the event from the database:

```
(venv)$ curl -X 'DELETE' \
  'http://0.0.0.0:8080/event/delete/1' \
  -H 'accept: application/json'
```

The request returns a successful response:

```
{
    "message": "Event deleted successfully"
}
```

Now, if we retrieve the list of events, we get an empty array for a response:

```
(venv)$ curl -X 'GET' \
  'http://0.0.0.0:8080/event/' \
  -H 'accept: application/json'
[]
```

We have successfully incorporated a SQL database into our application using SQLModel, as well as implementing CRUD operations. Let's commit the changes made to the application before learning how to implement CRUD operations in MongoDB:

```
(venv)$ git add .
(venv)$ git commit -m "[Feature] Incorporate a SQL database and
implement CRUD operations "
```

Switch back to the `main` branch:

```
(venv)$ git checkout main
```

Now that you are back to the original version of the application, let's incorporate MongoDB as the database platform and implement CRUD operations in the next section.

Setting up MongoDB

There are a number of libraries that allow us to integrate MongoDB into our FastAPI application. However, we'll be using **Beanie**, an asynchronous **Object Document Mapper (ODM)** library, to execute database operations from our application.

Let's install the `beanie` library by running the following command:

```
(venv)$ pip install beanie
```

Before diving into the integration, let's look at some of the methods from the Beanie library and also how database tables are created in this section.

Document

In SQL, the data stored in rows and columns are contained in the table. In a NoSQL database, it is called a document. The document represents how the data will be stored in the database collection. Documents are defined the same way a Pydantic model is defined, except that the `Document` class from the Beanie library is inherited instead.

An example document is defined as follows:

```
from beanie import Document

class Event(Document):
    name: str
    location: str

    class Settings:
        name = "events"
```

The `Settings` subclass is defined to tell the library to create the collection name passed in the MongoDB database.

Now that we know how to create a document, let's look at the methods used for carrying out CRUD operations:

- `.insert()` and `.create()`: The `.insert()` and `.create()` methods are called by the document instance to create a new record in the database. You can also choose to use the `.insert_one()` method to add a singular entry to the database.

  To insert many entries into the database, the `.insert_many()` method, which takes a list of the document instance, is called, such as the following:

```
event = Event(name="Packt office launch",
location="Hybrid")
await event.create()
await Event.insert_one(event)
```

- `.find()` and `.get()`: The `.find()` method is used to find a list of documents matching the search criteria passed as the method argument. The `.get()` method is used to retrieve a single document matching the supplied ID. A single document matching a search criterion can be found using the `.find_one()` method, such as the following:

```
event = await Event.get("74478287284ff")
event = await Event.find(Event.location == "Hybrid").to_
list() # Returns a list of matching items
event = await.find_one(Event.location == "Hybrid") #
Returns a single event
```

- `.save()`, `.update()`, and `.upsert()`: To update a document, any of these methods can be used. The `.update()` method takes an update query, and the `.upsert()` method is used when a document doesn't match the search criteria. In this chapter, we'll be making use of the `.update()` method. An update query is an instruction followed by the MongoDB database, such as the following:

```
event = await Event.get("74478287284ff")
update_query = {"$set": {"location": "virtual"}}
await event.update(update_query)
```

In this code block, we first retrieve the event and then create an update query to set the location field in the event collection to virtual.

- `.delete()`: This method is responsible for removing a document record from the database, such as the following:

```
event = await Event.get("74478287284ff")
await event.delete()
```

Now that we have learned how the methods contained in the Beanie library work, let's initialize the database in our event planner application, define our documents, and implement the CRUD operations.

Initializing the database

Let's look at the steps to do this:

1. In the database folder, create a `connection.py` file:

   (venv)$ touch connection.py

   Pydantic enables us to read environment variables by creating a child class of the `BaseSettings` parent class. When building web APIs, it is a standard practice to store configuration variables in an environment file.

2. In `connection.py`, add the following:

```
from beanie import init_beanie
from motor.motor_asyncio import AsyncIOMotorClient
from typing import Optional
from pydantic import BaseSettings


class Settings(BaseSettings):
    DATABASE_URL: Optional[str] = None

    async def initialize_database(self):
        client = AsyncIOMotorClient(self.DATABASE_URL)
        await init_beanie(
        database=client.get_default_database(),
                    document_models=[])

    class Config:
        env_file = ".env"
```

In this code block, we start by importing the dependencies required for initializing the database. Then, we define the Settings class, which has a DATABASE_URL value that is read from the env_file environment defined in the Config subclass. We also define an initialize_database method to initialize the database.

The init_beanie method takes the database client, which is the mongo version of the engine created in the SQLModel section, and a list of documents.

3. Let's update the model files in the models directory to include the MongoDB documents. In models/events.py, replace the contents with the following:

```python
from beanie import Document
from typing import Optional, List


class Event(Document):
    title: str
    image: str
    description: str
    tags: List[str]
    location: str


    class Config:
        schema_extra = {
            "example": {
                "title": "FastAPI Book Launch",
                "image": "https:
//linktomyimage.com/image.png",
                "description": "We will be discussing
                the contents of the FastAPI book in
                this event. Ensure to come with your
                own copy to win gifts!",
                "tags": ["python", "fastapi", "book",
                "launch"],
                "location": "Google Meet"
            }
        }
```

```
class Settings:
    name = "events"
```

4. Let's create a Pydantic model for the UPDATE operations:

```
class EventUpdate(BaseModel):
    title: Optional[str]
    image: Optional[str]
    description: Optional[str]
    tags: Optional[List[str]]
    location: Optional[str]

    class Config:
        schema_extra = {
            "example": {
                "title": "FastAPI Book Launch",
                "image": "https:
                //linktomyimage.com/image.png",
                "description": "We will be discussing
                the contents of the FastAPI book in
                this event. Ensure to come with your
                own copy to win gifts!",
                "tags": ["python", "fastapi", "book",
                "launch"],
                "location": "Google Meet"
            }
        }
```

5. In model/users.py, replace the content of the file with the following:

```
from typing import Optional, List
from beanie import Document, Link

from pydantic import BaseModel, EmailStr

from models.events import Event
```

```python
class User(Document):
    email: EmailStr
    password: str
    events: Optional[List[Link[Event]]]

    class Settings:
        name = "users"

    class Config:
        schema_extra = {
            "example": {
                "email": "fastapi@packt.com",
                "password": "strong!!!",
                "events": [],
            }
        }


class UserSignIn(BaseModel):
    email: EmailStr
    password: str
```

6. Now that we have defined the documents, let's update the document_models field in connection.py:

```python
from models.users import User
from models.events import Event

async def initialize_database(self):
        client = AsyncIOMotorClient(self.DATABASE_URL)
        await init_beanie(
        database=client.get_default_database(),
                        document_models=[Event, User])
```

7. Lastly, let's create an environment file, .env, and add the database URL to finalize the database initialization stage:

```
(venv)$ touch .env
(venv)$ echo DATABASE_URL=mongodb://localhost:27017/
planner >> .env
```

Now that we have successfully added the blocks of code to initialize the database, let's proceed to implement the methods for CRUD operations.

CRUD operations

In `connection.py`, create a new `Database` class that takes a model as an argument during initialization:

```
from pydantic import BaseSettings, BaseModel
from typing import Any, List, Optional


class Database:
    def __init__(self, model):
        self.model = model
```

The model passed during initialization is either the `Event` or `User` document model class.

Create

Let's create a method under the `Database` class to add a record to the database collection:

```
    async def save(self, document) -> None:
        await document.create()
            return
```

In this code block, we have defined the `save` method to take the document, which will be an instance of the document passed to the `Database` instance at the point of instantiation.

Read

Let's create the methods to retrieve a database record or all the records present in the database collection:

```
    async def get(self, id: PydanticObjectId) -> Any:
        doc = await self.model.get(id)
        if doc:
            return doc
```

```
    return False

async def get_all(self) -> List[Any]:
    docs = await self.model.find_all().to_list()
    return docs
```

The first method, get(), takes an ID as the method argument and returns
a corresponding record from the database, while the get_all() method takes no
argument and returns a list of all the records present in the database.

Update

Let's create the method to handle the process of updating an existing record:

```
async def update(self, id: PydanticObjectId, body:
BaseModel) -> Any:
    doc_id = id
    des_body = body.dict()
    des_body = {k:v for k,v in des_body.items() if v is
    not None}
    update_query = {"$set": {
        field: value for field, value in
        des_body.items()
    }}

    doc = await self.get(doc_id)
    if not doc:
        return False
    await doc.update(update_query)
    return doc
```

In this code block, the update method takes an ID and the Pydantic schema responsible,
which will contain the fields updated from the PUT request sent by the client. The updated
request body is first parsed into a dictionary and then filtered to remove None values.
Once this has been done, it is then inserted into an update query, which is finally executed
by Beanie's update() method.

Delete

Lastly, let's create a method to delete a record from the database:

```
async def delete(self, id: PydanticObjectId) -> bool:
    doc = await self.get(id)
    if not doc:
        return False
    await doc.delete()
    return True
```

In this code block, the method checks whether such a record exists before proceeding to delete it from the database.

Now that we have populated our database file with the necessary methods needed to carry out CRUD operations, let's update the routes as well.

routes/events.py

Let's start by updating the imports and creating a database instance:

```
from beanie import PydanticObjectId
from fastapi import APIRouter, HTTPException, status
from database.connection import Database

from models.events import Event
from typing import List
event_database = Database(Event)
```

With the imports and database instance in place, let's update all the routes. Start by updating the GET routes:

```
@event_router.get("/", response_model=List[Event])
async def retrieve_all_events() -> List[Event]:
    events = await event_database.get_all()
    return events


@event_router.get("/{id}", response_model=Event)
async def retrieve_event(id: PydanticObjectId) -> Event:
    event = await event_database.get(id)
    if not event:
```

```
        raise HTTPException(
            status_code=status.HTTP_404_NOT_FOUND,
            detail="Event with supplied ID does not exist"
        )
    return event
```

In the GET routes, we are invoking the methods we defined in the database file earlier. Let's update the POST routes:

```
@event_router.post("/new")
async def create_event(body: Event) -> dict:
    await event_database.save(body)
    return {
        "message": "Event created successfully"
    }
```

Let's create the UPDATE route:

```
@event_router.put("/{id}", response_model=Event)
async def update_event(id: PydanticObjectId, body: EventUpdate)
-> Event:
    updated_event = await event_database.update(id, body)
    if not updated_event:
        raise HTTPException(
            status_code=status.HTTP_404_NOT_FOUND,
            detail="Event with supplied ID does not exist"
        )
    return updated_event
```

Lastly, let's update the DELETE route:

```
@event_router.delete("/{id}")
async def delete_event(id: PydanticObjectId) -> dict:
    event = await event_database.delete(id)
    if not event:
        raise HTTPException(
            status_code=status.HTTP_404_NOT_FOUND,
            detail="Event with supplied ID does not exist"
        )
```

```
    return {
        "message": "Event deleted successfully."
    }
```

Now that we have implemented the CRUD operations to our event routes, let's implement the routes for signing a user up and signing a user in.

routes/users.py

Let's start by updating the imports and creating a database instance:

```
from fastapi import APIRouter, HTTPException, status
from database.connection import Database

from models.users import User, UserSignIn

user_router = APIRouter(
    tags=["User"],
)


user_database = Database(User)
```

Next, update the POST route for signing new users:

```
@user_router.post("/signup")
async def sign_user_up(user: User) -> dict:
    user_exist = await User.find_one(User.email ==
    user.email)
    if user_exist:
        raise HTTPException(
            status_code=status.HTTP_409_CONFLICT,
            detail="User with email provided exists
            already."
        )
    await user_database.save(user)
    return {
        "message": "User created successfully"
    }
```

In this code block, we check whether such a user with the email passed exists before adding them to the database. Let's add the route to sign users in:

```
@user_router.post("/signin")
async def sign_user_in(user: UserSignIn) -> dict:
    user_exist = await User.find_one(User.email ==
    user.email)
    if not user_exist:
        raise HTTPException(
            status_code=status.HTTP_404_NOT_FOUND,
            detail="User with email does not exist."
        )
    if user_exist.password == user.password:
        return {
            "message": "User signed in successfully."
        }
    raise HTTPException(
        status_code=status.HTTP_401_UNAUTHORIZED,
        detail="Invalid details passed."
    )
```

In this defined route, we first check whether the user exists before checking the validity of their credentials. The method of authentication used here is basic and *not recommended* in production. We'll take a look at proper authentication procedures in the next chapter.

Now that we have implemented the routes, let's start a MongoDB instance as well as our application. Create a folder to house our MongoDB database and start the MongoDB instance:

```
(venv)$ mkdir store
(venv)$ mongod --dbpath store
```

Next, in another window, start the application:

```
(venv)$ python main.py
INFO:     Uvicorn running on http://0.0.0.0:8080 (Press CTRL+C
to quit)
INFO:     Started reloader process [3744] using statreload
INFO:     Started server process [3747]
INFO:     Waiting for application startup.
INFO:     Application startup complete.
```

Let's test the event routes:

1. Create an event:

```
(venv)$ curl -X 'POST' \
  'http://0.0.0.0:8080/event/new' \
  -H 'accept: application/json' \
  -H 'Content-Type: application/json' \
  -d '{
  "title": "FastAPI Book Launch",
  "image": "https://linktomyimage.com/image.png",
  "description": "We will be discussing the contents
  of the FastAPI book in this event. Ensure to come
  with your own copy to win gifts!",
  "tags": [
    "python",
    "fastapi",
    "book",
    "launch"
  ],
  "location": "Google Meet"
}'
```

   Here is the response from the preceding operation:

```
{
  "message": "Event created successfully"
}
```

2. Get all events:

```
(venv)$ curl -X 'GET' \
  'http://0.0.0.0:8080/event/' \
  -H 'accept: application/json'
```

The preceding request returns a list of events:

```
[
    {
        "_id": "624daab1585059e8a3fa77ac",
        "title": "FastAPI Book Launch",
        "image": "https://linktomyimage.com/image.png",
        "description": "We will be discussing the contents
        of the FastAPI book in this event. Ensure to come
        with your own copy to win gifts!",
        "tags": [
          "python",
          "fastapi",
          "book",
          "launch"
        ],
        "location": "Google Meet"
    }
]
```

3. Get an event:

```
(venv)$ curl -X 'GET' \
  'http://0.0.0.0:8080/event/624daab1585059e8a3fa77ac' \
  -H 'accept: application/json'
```

This operation returns the event that matches the supplied ID:

```
{
    "_id": "624daab1585059e8a3fa77ac",
    "title": "FastAPI Book Launch",
    "image": "https://linktomyimage.com/image.png",
    "description": "We will be discussing the contents
    of the FastAPI book in this event. Ensure to come
    with your own copy to win gifts!",
```

```
  "tags": [
    "python",
    "fastapi",
    "book",
    "launch"
  ],
  "location": "Google Meet"
}
```

4. Let's update the event location to `Hybrid`:

```
(venv)$ curl -X 'PUT' \
  'http://0.0.0.0:8080/event/624daab1585059e8a3fa77ac'
  \
  -H 'accept: application/json' \
  -H 'Content-Type: application/json' \
  -d '{
  "location": "Hybrid"
}'

{
  "_id": "624daab1585059e8a3fa77ac",
  "title": "FastAPI Book Launch",
  "image": "https://linktomyimage.com/image.png",
  "description": "We will be discussing the contents
  of the FastAPI book in this event. Ensure to come
  with your own copy to win gifts!",
  "tags": [
    "python","fastapi",
    "book",
    "launch"
  ],
  "location": "Hybrid"
}
```

5. Lastly, let's delete the event:

```
(venv)$ curl -X 'DELETE' \
  'http://0.0.0.0:8080/event/624daab1585059e8a3fa77ac'
  \
  -H 'accept: application/json'
```

Here is the response received to the request:

```
{
    "message": "Event deleted successfully."
}
```

6. Now that we have tested the routes for the events, let's create a new user and then
    sign in:

```
(venv)$ curl -X 'POST' \
  'http://0.0.0.0:8080/user/signup' \
  -H 'accept: application/json' \
  -H 'Content-Type: application/json' \
  -d '{
  "email": "fastapi@packt.com",
  "password": "strong!!!",
  "events": []
}'
```

The request returns a response:

```
{
    "message": "User created successfully"
}
```

Running the request again returns an HTTP 409 error, indicating a conflict:

```
{
    "detail": "User with email provided exists already."
}
```

We originally designed the route to check for existing users to avoid duplicates.

7. Now, let's send a POST request to sign in the user we just created:

```
(venv)$ curl -X 'POST' \
  'http://0.0.0.0:8080/user/signin' \
  -H 'accept: application/json' \
  -H 'Content-Type: application/json' \
  -d '{
  "email": "fastapi@packt.com",
  "password": "strong!!!"
}'
```

The request returns an HTTP 200 success message:

```
{
    "message": "User signed in successfully."
}
```

We have successfully implemented CRUD operations using the Beanie library.

Summary

In this chapter, we learned how to add SQL and NoSQL databases using SQLModel and Beanie respectively. We made use of all our knowledge from the previous chapters. We also tested the routes to ensure that they are working as planned.

In the next chapter, you will be introduced to securing your application. You will first be taught the basics of authentication as well as the various authentication methods available to FastAPI developers. You will then implement an authentication system that relies on **JSON Web Token** (**JWT**) and secure the routes to create, update, and delete events. Lastly, you will modify the route, to create events to allow the linking of events to a user.

7
Securing FastAPI Applications

In the last chapter, we looked at how to connect a FastAPI application to a SQL and NoSQL database. We successfully implemented database methods and updated the existing routes to enable interactions between the application and the database. However, the planner application continues to allow anybody to add an event as opposed to only authenticated users. In this chapter, we will secure the application using **JSON Web Token (JWT)** and restrict some event operations to only authenticated users.

Securing an application involves the addition of security measures to restrict access to application functionalities from unauthorized entities to prevent hacks or illegal modifications of the application. Authentication is the process of verifying the credentials passed by an entity and authorization simply means giving an entity permission to perform designated actions. When credentials have been verified, the entity is then authorized to carry out various actions.

By the end of this chapter, you will be able to add an authentication layer to a FastAPI application. This chapter will explain the processes for securing passwords by hashing them, adding an authentication layer, and securing routes from unauthenticated users. In this chapter, we'll be covering the following topics:

- Authentication methods in FastAPI

- Securing the application with OAuth2 and JWT

- Protecting routes using dependency injection

- Configuring CORS

Technical requirements

To follow along, the MongoDB database component is required. The installation procedures for your operating system can be found in their official documentation. The code used in this chapter can be found at `https://github.com/PacktPublishing/Building-Python-Web-APIs-with-FastAPI/tree/main/ch07/planner`.

Authentication methods in FastAPI

There are several authentication methods available in FastAPI. FastAPI supports the common authentication methods of basic HTTP authentication, cookies, and bearer token authentication. Let's briefly look at what each method entails:

- **Basic HTTP authentication**: In this authentication method, the user credentials, which is usually a username and password, are sent via an `Authorization` HTTP header. The request in turn returns a `WWW-Authenticate` header containing a `Basic` value and an optional realm parameter, which indicates the resource the authentication request is made to.

- **Cookies**: Cookies are employed when data is to be stored on the client side, such as in web browsers. FastAPI applications can also employ cookies to store user data, which can be retrieved by the server for authentication purposes.

- **Bearer token authentication**: This method of authentication involves the use of security tokens called bearer tokens. These tokens are sent alongside the `Bearer` keyword in an `Authorization` header request. The most used token is JWT, which is usually a dictionary comprising the user ID and the token's expiry time.

Every authentication method listed here has its specific use cases as well as its pros and cons. However, in this chapter, we'll be making use of bearer token authentication.

Authentication methods are injected into FastAPI applications as dependencies that are called at runtime. This simply means when authentication methods are defined, they are dormant until injected into their place of use. This activity is called **Dependency Injection**.

Dependency injection

Dependency injection is a pattern where an object – in this case, a function – receives an instance variable needed for the further execution of the function.

In FastAPI, dependencies are injected by declaring them in the path operation function arguments. We have been using dependency injection in previous chapters. Here's an example from the previous chapter where we retrieve the email field from the user model passed to the function:

```
@user_router.post("/signup")
async def sign_user_up(user: User) -> dict:
    user_exist = await User.find_one(User.email == user.email)
```

In this code block, the dependency defined is the User model class, which is injected into the sign_user_up() function. By injecting the User model into the user function argument, we can easily retrieve the attributes of the object.

Creating and using a dependency

In FastAPI, a dependency can be defined as either a function or a class. The dependency created gives us access to its underlying values or methods, eliminating the need to create these objects in the functions inheriting them. Dependency injection helps in reducing code repetition in some cases, such as in enforcing authentication and authorization.

An example dependency is defined as follows:

```
async def get_user(token: str):
    user = decode_token(token)
    return user
```

This dependency is a function that takes token as the argument and returns a user parameter from an external function, decode_token. To use this dependency, the dependent function argument declared is set to have a Depends parameter, for example:

```
from fastapi import Depends


@router.get("/user/me")
```

```
async get_user_details(user: User = Depends(get_user)):
    return user
```

The route function here is dependent on the get_user function, which serves as its dependency. What this means is that to access the preceding route, the get_user dependency must be satisfied.

The Depends class, which is imported from the FastAPI library, is responsible for taking the function passed as the argument and executing it when the endpoint is called, automatically making available to the endpoint, they return value of the function passed to it.

Now that you have an idea of how a dependency is created and how it's used, let's build the authentication dependency for the event planner application.

Securing the application with OAuth2 and JWT

In this section, we'll build out the authentication system for the event planner application. We'll be making use of the OAuth2 password flow, which requires the client to send a username and password as form data. The username in our case is the email used when creating an account.

When the form data is sent to the server from the client, an **access token**, which is a signed JWT, is sent as a response. Usually, a background check is done to validate the credentials sent to the server before creating a token to allow further authorization. To authorize the authenticated user, the JWT is prefixed with Bearer when sent via the header to authorize the action on the server.

> **What Is a JWT and Why Is It Signed?**
>
> A JWT is an encoded string usually containing a dictionary housing a payload, a signature, and its algorithm. JWTs are signed using a unique key known only to the server and client to avoid the encoded string being tampered with by an external body.

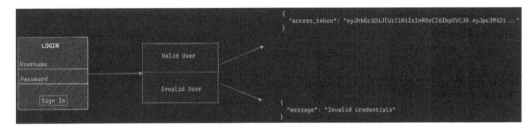

Figure 7.1 – Authentication flow

Now that we have an idea of how the authentication flow works, let's create the necessary folder and files required to set up an authentication system in our application:

1. In the project folder, create the auth folder first:

   ```
   (venv)$ mkdir auth
   ```

2. Next, create the following files in the auth folder:

   ```
   (venv)$ cd auth && touch {__init__,jwt_
   handler,authenticate,hash_password}.py
   ```

The preceding command creates four files:

- jwt_handler.py: This file will contain the functions required to encode and decode the JWT strings.
- authenticate.py: This file will contain the authenticate dependency, which will be injected into our routes to enforce authentication and authorization.
- hash_password.py: This file will contain the functions that will be used to encrypt the password of a user during sign-up and compare passwords during sign-in.
- __init__.py: This file indicates the contents of the folder as a module.

Now that the files have been created, let's build the individual components. We'll start by creating the components for hashing user passwords.

Hashing passwords

In the previous chapter, we stored user passwords in plain text. This is a highly insecure and prohibited practice when building APIs. Passwords are to be encrypted or hashed using appropriate libraries. We'll be encrypting the user passwords using bcrypt.

Let's install the passlib library. This library houses the bcrypt hashing algorithm, which we'll be using for hashing user passwords:

```
(venv)$ pip install passlib[bcrypt]
```

Now that we have installed the library, let's create the functions for hashing the passwords in hash_password.py:

```
from passlib.context import CryptContext


pwd_context = CryptContext(schemes=["bcrypt"],
```

```
deprecated="auto")

class HashPassword:
    def create_hash(self, password: str):
        return pwd_context.hash(password)

    def verify_hash(self, plain_password: str,
    hashed_password: str):
        return pwd_context.verify(plain_password,
        hashed_password)
```

In the preceding code block, we start by importing CryptContext, which takes the bcrypt scheme for hashing the strings passed to it. The context is stored in the pwd_context variable, giving us access to the methods required for executing our task.

The HashPassword class is then defined and contains two methods, create_hash and verify_hash:

- The create_hash method takes a string and returns the hashed value.

- verify_hash takes the plain password and the hashed password and compares them. The function returns a Boolean value indicating whether the values passed are the same or not.

Now that we have created a class to handle the hashing of passwords, let's update the sign-up route to hash the user password before storing it in the database:

routes/users.py

```
from auth.hash_password import HashPassword
from database.connection import Database

user_database = Database(User)
hash_password = HashPassword()

@user_router.post("/signup")
async def sign_user_up(user: User) -> dict:
    user_exist = await User.find_one(User.email ==
```

```
user.email)

if user_exist:
    raise HTTPException(
        status_code=status.HTTP_409_CONFLICT,
        detail="User with email provided exists
        already."
    )
hashed_password = hash_password.create_hash(
user.password)
user.password = hashed_password
await user_database.save(user)
return {
    "message": "User created successfully"
}
```

Now that we have updated the user sign-up route to hash the password before saving, let's create a new user to confirm. In a terminal window, start the application:

```
(venv)$ python main.py
INFO:      Uvicorn running on http://0.0.0.0:8080 (Press CTRL+C
to quit)
INFO:      Started reloader process [8144] using statreload
INFO:      Started server process [8147]
INFO:      Waiting for application startup.
INFO:      Application startup complete.
```

In another terminal window, start the MongoDB instance:

```
$ mongod --dbpath database --port 27017
```

Next, let's create a new user:

```
$ curl -X 'POST' \
  'http://0.0.0.0:8080/user/signup' \
  -H 'accept: application/json' \
  -H 'Content-Type: application/json' \
  -d '{
  "email": "reader@packt.com",
```

```
    "password": "exemplary"
}'
```

We get a success response from the request above:

```
{
    "message": "User created successfully"
}
```

Now that we have created a user, let's verify that the password sent to the database was hashed. To do that, we'll create an interactive MongoDB session that allows us to run commands from within the database.

In a new terminal window, run the following commands:

```
$ mongo --port 27017
```

An interactive MongoDB session is started:

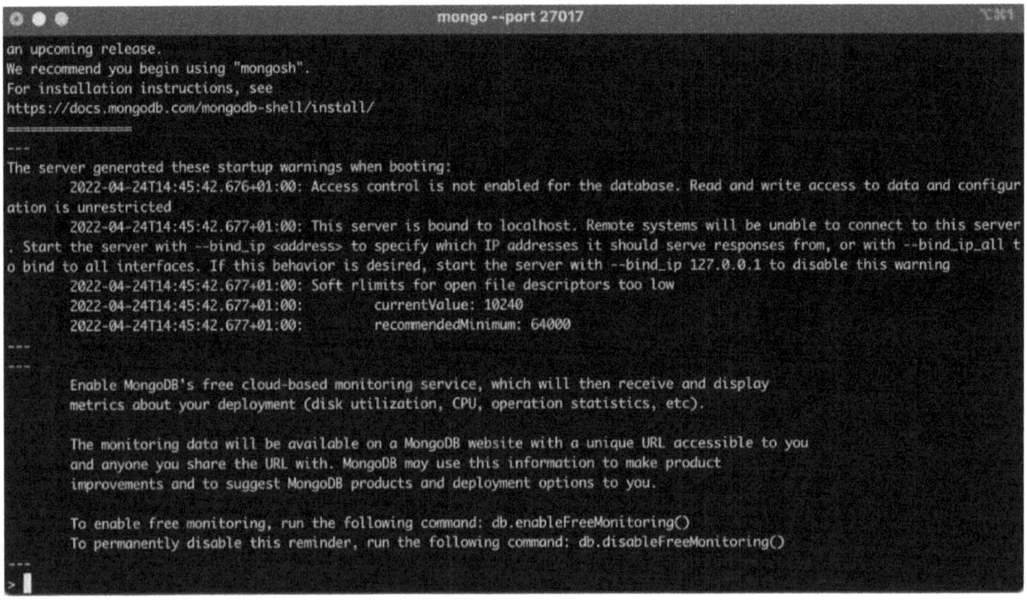

Figure 7.2 – An interactive MongoDB session

With the interactive session running, run the series of commands to switch to the planner database and retrieve all user records:

```
> use planner
> db.users.find({})
```

> use planner
switched to db planner
> db.users.find({})
{ "_id" : ObjectId("62655d4b52b6386b8b11b5fb"), "email" : "reader@packt.com", "password" : "$2b$12$Jcc5VXty397UDGeg3bdq0encodqNvi
f8npVjO6P1IU1NfIjONGP/m", "events" : [ ] }

Figure 7.3 – Result from the find all users query

The preceding command returns the list of users, and we can now confirm that the user's password was hashed before it was stored in the database. Now that we have successfully built the components for securely storing user passwords, let's build the components for creating and verifying JWTS.

Creating and verifying access tokens

Creating a **JWT takes us** a step closer to securing our application. The token's payload will comprise the user ID and an expiry time before encoding in the long string as shown here:

```
{
  "access_token": "eyJhbGciOiJIUzI1NiIsInR5cCI6IkpXVCJ9.eyJpc3MiOi ... "
}
```

User	62655d4b52b6386b8b11b5fb
Expires	datetime.datetime(2022, 4, 24, 15, 39, 16, 630218)

Figure 7.4 – Anatomy of a JWT

Earlier on, we learned why JWTs are signed. JWTs are signed with a secret key known only to the sender and the receiver. Let's update the Settings class in database/ database.py as well as the environment file, .env, to include a SECRET_KEY variable, which will be used to sign the JWTs:

database/database.py

```
class Settings(BaseSettings):
    SECRET_KEY: Optional[str] = None
```

.env

```
SECRET_KEY=HI5HL3V3L$3CR3T
```

With that in place, add the following imports in jwt_handler.py:

```
import time
from datetime import datetime

from fastapi import HTTPException, status
from jose import jwt, JWTError
from database.database import Settings
```

In the preceding code block, we have imported the time modules, the HTTPException class, as well as the status from FastAPI. We also imported the jose library responsible for encoding and decoding JWTs and the Settings class.

Next, we'll create an instance of the Settings class so we can retrieve the SECRET_KEY variable and create the function responsible for creating the token:

```
settings = Settings()

def create_access_token(user: str) -> str:
    payload = {
        "user": user,
        "expires": time.time() + 3600
    }

    token = jwt.encode(payload, settings.SECRET_KEY,
    algorithm="HS256")
    return token
```

In the preceding code block, the function takes a string argument, which is passed into the payload dictionary. The payload dictionary contains the user and the expiry time, which is returned when a JWT is decoded.

The `expires` value is set to an hour from the time of creation. The payload is then passed to the `encode()` method, which takes three parameters:

- **Payload**: A dictionary containing the values to be encoded.

- **Key**: The key used to sign the payload.

- **Algorithm**: The algorithm used in signing the payload. The default and most common is the **HS256** algorithm.

Next, let's create a function to verify the authenticity of a token sent to our application:

```python
def verify_access_token(token: str) -> dict:
    try:
        data = jwt.decode(token, settings.SECRET_KEY,
        algorithms=["HS256"])

        expire = data.get("expires")

        if expire is None:
            raise HTTPException(
                status_code=status.HTTP_400_BAD_REQUEST,
                detail="No access token supplied"
            )
        if datetime.utcnow() >
        datetime.utcfromtimestamp(expire):
            raise HTTPException(
                status_code=status.HTTP_403_FORBIDDEN,
                detail="Token expired!"
            )
        return data

    except JWTError:
        raise HTTPException(
            status_code=status.HTTP_400_BAD_REQUEST,
            detail="Invalid token"
        )
```

In the preceding code block, the function takes the token string as the argument and runs several checks in the `try` block. The function first checks the expiry time of the token. If there's no expiry time, then no token was supplied. The second check is the validity of the token – an exception is thrown to inform the user of the token expiration. If the token is valid, the decoded payload is returned.

In the `except` block, a bad request exception is thrown for any JWT error.

Now that we have implemented the functions for creating and verifying the tokens sent to the application, let's create the function that validates user authentication and serves as the dependency.

Handling user authentication

We have successfully implemented the components for hashing and comparing passwords as well as components for creating and decoding JWTs. Let's implement the dependency function that will be injected into the event routes. This function will serve as the single source of truth for retrieving a user for an active session.

In auth/authenticate.py, add the following:

```
from fastapi import Depends, HTTPException, status
from fastapi.security import OAuth2PasswordBearer

from auth.jwt_handler import verify_access_token

oauth2_scheme = OAuth2PasswordBearer(tokenUrl="/user/signin")

async def authenticate(token: str = Depends(oauth2_scheme)) ->
str:
    if not token:
        raise HTTPException(
            status_code=status.HTTP_403_FORBIDDEN,
            detail="Sign in for access"
        )

    decoded_token = verify_access_token(token)
    return decoded_token["user"]
```

In the preceding code block, we start by importing the necessary dependencies:

- `Depends`: This injects `oauth2_scheme` into the function as a dependency.
- `OAuth2PasswordBearer`: This class tells the application that a security scheme is present.
- `verify_access_token`: This function, defined in the creating and verifying access token section will be used to check the validity of the token.

We then define the token URL for the OAuth2 scheme and the `authenticate` function. The `authenticate` function takes the token as the argument. The function has the OAuth scheme injected into it as a dependency. The token is decoded, and the user field of the payload is returned if the token is valid, otherwise, the adequate error responses are returned as defined in the `verify_access_token` function.

Now that we have successfully created the dependency for securing the routes, let's update the authentication flow in the routes, as well as injecting the `authenticate` function into the event routes.

Updating the application

In this section, we'll update the routes to use the new authentication model. Lastly, we'll update the POST route for adding an event to populate the events field in the user's record.

Updating the user sign-in route

In `routes/users.py`, update the imports:

```
from fastapi import APIRouter, Depends, HTTPException, status
from fastapi.security import OAuth2PasswordRequestForm
from auth.jwt_handler import create_access_token


from models.users import User
```

We have imported the `OAuth2PasswordRequestForm` class from FastAPI's security module. This will be injected into the sign-in route to retrieve the credentials sent over: username and password. Let's update the `sign_user_in()` route function:

```
async def sign_user_in(user: OAuth2PasswordRequestForm =
Depends()) -> dict:
    user_exist = await User.find_one(User.email ==
```

```
user.username)

. .

if hash_password.verify_hash(user.password,
user_exist.password):
    access_token = create_access_token(
    user_exist.email)
    return {
        "access_token": access_token,
        "token_type": "Bearer"
    }
```

In the preceding code block, we have injected the `OAuth2PasswordRequestForm` class as the dependency for this function, ensuring the OAuth spec is strictly followed. In the function body, we compare the password and return an access token and a token type. Before we test the updated route, let's create a response model for the login route in `models/users.py` to replace the `UserSignIn` model class, which isn't used anymore:

```
class TokenResponse(BaseModel):
    access_token: str
    token_type: str
```

Update the imports and the response model for the sign-in route:

```
from models.users import User, TokenResponse
```

```
@user_router.post("/signin", response_model=TokenResponse)
```

Let's visit the interactive docs to confirm that the request body is compliant with the OAuth2 specs at `http://0.0.0.0:8080/docs`:

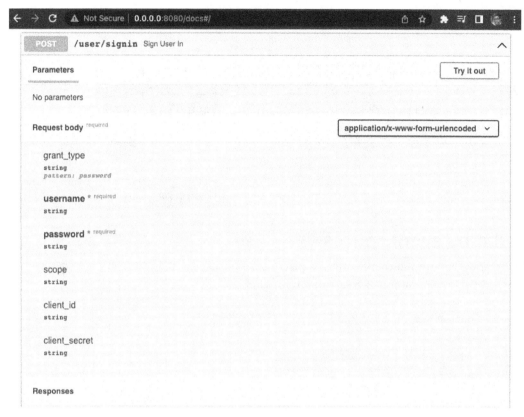

Figure 7.5 – Request body for updated sign-in route

Let's sign in to verify that the route works properly:

```
$ curl -X 'POST' \
  'http://0.0.0.0:8080/user/signin' \
  -H 'accept: application/json' \
  -H 'Content-Type: application/x-www-form-urlencoded' \
  -d 'grant_type=&username=reader%40packt.
com&password=exemplary&scope=&client_id=&client_secret='
```

The response returned is an access token and the token type:

```
{
  "access_token": "eyJhbGciOiJIUzI1NiIsInR5cCI6IkpXVCJ9.eyJ1c2
VyIjoicmVhZGVyQHBhY2t0LmNvbSIsImV4cGlyZXMiOjE2NTA4Mjc0MjQuMDg2
NDAxfQ.LY4i5EjIzlsKdfMyWKi7XH7lLeDuVt3832hNfkQx8C8",
  "token_type": "Bearer"
}
```

Now that we have confirmed that the route works as expected, let's update the event routes to allow only authorized users' **CREATE**, **UPDATE**, and **DELETE** events.

Updating event routes

Now that we have our authentication in place, let's inject the authentication dependency into the POST, PUT, and DELETE route functions:

```
from auth.authenticate import authenticate

async def create_event(body: Event, user: str =
Depends(authenticate)) -> dict:
    ..


async def update_event(id: PydanticObjectId, body: EventUpdate,
user: str = Depends(authenticate)) -> Event:
    ..


async def delete_event(id: PydanticObjectId, user: str =
Depends(authenticate)) -> dict:
    ..
```

With the dependencies injected, the interactive docs website is automatically updated to show protected routes. If we log on to http://0.0.0.0:8080/docs, we can see the **Authorize** button at the top right and the padlocks on the event routes:

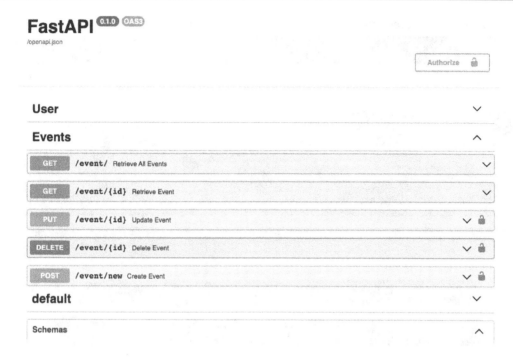

Figure 7.6 – Updated documentation page

If we click on the **Authorize** button, a sign-in modal is displayed. Inputting our credentials and password returns the following screen:

Figure 7.7 – Authenticated user

Now that we have successfully signed in, we can create an event:

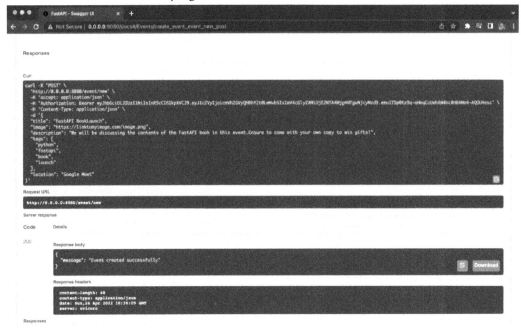

Figure 7.8 – Create a new event

The same operations can be performed from the command line. First, let's get our access token:

```
$ curl -X 'POST' \
  'http://0.0.0.0:8080/user/signin' \
  -H 'accept: application/json' \
  -H 'Content-Type: application/x-www-form-urlencoded' \
  -d 'grant_type=&username=reader%40packt.
com&password=exemplary&scope=&client_id=&client_secret='
```

The request sent returns the access token, which is a JWT string, and the token type, which is of type `Bearer`:

```
{
  "access_token": "eyJhbGciOiJIUzI1NiIsInR5cCI6IkpXVCJ9.
eyJlc2VyIjoicmVhZGVyQHBhY2t0LmNvbSIsImV4cGlyZXMiOjE2NTA4MjkxOD
MuNTg3NjAyfQ.MOXjI5GXnyzGNftdlxDGyM119_L11uPq8yCxBHepf04",
  "token_type": "Bearer"
}
```

Now, let's create a new event from the command line:

```
$ curl -X 'POST' \
   'http://0.0.0.0:8080/event/new' \
   -H 'accept: application/json' \
   -H 'Authorization: Bearer eyJhbGciOiJIUzI1NiIsInR5cCI6IkpXV
CJ9.eyJlc2VyIjoicmVhZGVyQHBhY2t0LmNvbSIsImV4cGlyZXMiOjE2NTA4Mjk
xODMuNTg3NjAyfQ.MOXjI5GXnyzGNftdlxDGyM119_L1luPq8yCxBHepf04' \
   -H 'Content-Type: application/json' \
   -d '{
   "title": "FastAPI Book Launch CLI",
   "image": "https://linktomyimage.com/image.png",
   "description": "We will be discussing the contents of the
   FastAPI book in this event.Ensure to come with your own
   copy to win gifts!",
   "tags": [
      "python",
      "fastapi",
      "book",
      "launch"
   ],
   "location": "Google Meet"
}'
```

In the request sent here, the `Authorization: Bearer` header is sent as well to inform the application that we are authorized to perform this action. The response gotten is the following:

```
{ "message": "Event created successfully" }
```

If we try to create an event without passing the authorization header with a valid token, an `HTTP 401 Unauthorized` error is returned:

```
$ curl -X 'POST' \
   'http://0.0.0.0:8080/event/new' \
   -H 'accept: application/json' \
   -H 'Content-Type: application/json' \
   -d '{
   "title": "FastAPI BookLaunch",
```

```
    "image": "https://linktomyimage.com/image.png",
    "description": "We will be discussing the contents of the
    FastAPI book in this event.Ensure to come with your own
    copy to win gifts!",
    "tags": [
      "python",
      "fastapi",
      "book",
      "launch"
    ],
    "location": "Google Meet"
  }'
```

Here's the response:

```
$ {
  "detail": "Not authenticated"
}
```

Now that we have successfully protected the routes, let's update the protected routes as follows:

- POST route: Add the event created to the list of events owned by the user.

- UPDATE route: Modify the route to ensure only the event created by the user can be updated.

- DELETE route: Modify the route to ensure only the event created by the user can be deleted.

In the previous section, we successfully injected the authentication dependencies to our route operations. To easily identify events and prevent a user from deleting another user's event, we'll update the event document class as well as the routes.

Updating the event document class and routes

Add the `creator` field to the `Event` document class in `models/events.py`:

```
class Event(Document):
    creator: Optional[str]
```

This field will enable us to restrict the operations performed on an event to the user alone.

Next, let's modify the POST route to update the `creator` field when creating a new event in `routes/events.py`:

```
@event_router.post("/new")
async def create_event(body: Event, user: str =
Depends(authenticate)) -> dict:
    body.creator = user
    await event_database.save(body)
    return {
        "message": "Event created successfully"
    }
```

In the preceding code block, we have updated the POST route to add the current user's email as the creator of the event. If you create a new event, the event is stored with the creator's email:

```
$ curl -X 'POST' \
  'http://0.0.0.0:8080/event/new' \
  -H 'accept: application/json' \
  -H 'Authorization: Bearer eyJhbGciOiJIUzI1NiIsInR5cCI6IkpXV
CJ9.eyJlc2VyIjoicmVhZGVyQHBhY2t0LmNvbSIsImV4cGlyZXMiOjE2NTA4MzI
5NjQuMTU3MjQ4fQ.RxR1TYMx91JtVMNzYcT7718xXWX7skTCfWbnJxyf6fU' \
  -H 'Content-Type: application/json' \
  -d '{
"title": "FastAPI Book Launch",
"image": "https://linktomyimage.com/image.png",
"description": "We will be discussing the contents of the
FastAPI book in this event.Ensure to come with your own
copy to win gifts!",
"tags": [
  "python",
  "fastapi",
  "book",
  "launch"
],
"location": "Google Meet"
```

```
}'
```

The response returned from the request above is:

```
{
  "message": "Event created successfully"
}
```

Next, let's retrieve the list of events stored in the database:

```
$ curl -X 'GET' \
  'http://0.0.0.0:8080/event/' \
  -H 'accept: application/json'
```

The response from the request above is:

```
[
  {
    "_id": "6265a807e0c8daefb72261ea",
    "creator": "reader@packt.com",
    "title": "FastAPI BookLaunch",
    "image": "https://linktomyimage.com/image.png",
    "description": "We will be discussing the contents of the
    FastAPI book in this event.Ensure to come with your own
    copy to win gifts!",
    "tags": [
      "python",
      "fastapi",
      "book",
      "launch"
    ],
    "location": "Google Meet"
  },
]
```

Next, let's update the UPDATE route:

```
@event_router.put("/{id}", response_model=Event)
async def update_event(id: PydanticObjectId, body: EventUpdate,
```

```
user: str = Depends(authenticate)) -> Event:
    event = await event_database.get(id)
    if event.creator != user:
        raise HTTPException(
            status_code=status.HTTP_400_BAD_REQUEST,
            detail="Operation not allowed"
        )
```

In the preceding code block, the route function checks whether the current user can edit an event before proceeding, otherwise, it raises an HTTP 400 bad request exception. Here's an example using a different user:

```
$ curl -X 'PUT' \
  'http://0.0.0.0:8080/event/6265a83fc823a3c912830074' \
  -H 'accept: application/json' \
  -H 'Authorization: Bearer eyJhbGciOiJIUzI1NiIsInR5cCI6IkpXVC
J9.eyJlc2VyIjoiZmFzdGFwaUBwYWNrdC5jb20iLCJleHBpcmVzIjoxNjUwODM
zOTc2LjI2NgzMX0.MMRT6pwEDBVHTU5C1a6MV8j9wCfWhqbza9NBpZz08xE' \
  -H 'Content-Type: application/json' \
  -d '{
  "title": "FastAPI Book Launch"
}'
```

Here's the response:

```
{
    "detail": "Operation not allowed"
}
```

Lastly, let's update the DELETE route:

```
@event_router.delete("/{id}")
async def delete_event(id: PydanticObjectId, user: str =
Depends(authenticate)):
    event = await event_database.get(id)
    if event.creator != user:
        raise HTTPException(
```

```
            status_code=status.HTTP_404_NOT_FOUND,
            detail="Event not found"
        )
```

In the preceding code block, we instruct the route function to first check whether the current user is the creator, otherwise raise an exception. Let's take a look at an example where another user attempts to delete another user's event:

```
$ curl -X 'DELETE' \
  'http://0.0.0.0:8080/event/6265a83fc823a3c912830074' \
  -H 'accept: application/json' \
  -H 'Authorization: Bearer eyJhbGciOiJIUzI1NiIsInR5cCI6IkpXV
CJ9.eyJlc2VyIjoiZmFzdGFwaUBwYWNrdC5jb20iLCJleHBpcmVzIjoxNjUwOD
MzOTc2LjI2NzgzMX0.MMRT6pwEDBVHTU5C1a6MV8j9wCfWhqbza9NBpZz08xE'
```

An event not found is returned as the response:

```
{
  "detail": "Event not found"
}
```

However, the ideal owner can delete an event:

```
$ curl -X 'DELETE' \
  'http://0.0.0.0:8080/event/6265a83fc823a3c912830074' \
  -H 'accept: application/json' \
  -H 'Authorization: Bearer eyJhbGciOiJIUzI1NiIsInR5cCI6IkpXVC
J9.eyJlc2VyIjoicmVhZGVyQHBhY2t0LmNvbSIsImV4cGlyZXMiOjE2NTA4MzQz
OTUuMDkzMDI3fQ.IKYHWQ2YO3rQc-KR8kyfoy_54MsEVE75WbRqoVbdoW0'
```

Here's the response:

```
{
  "message": "Event deleted successfully."
}
```

We have successfully secured our application and its routes. Let's wrap up this chapter by configuring a Cross-Origin Resource Sharing (CORS) middleware in the next section.

Configuring CORS

Cross-Origin Resource Sharing (CORS) serves as a rule that prevents unregistered clients access to a resource.

When our web API is consumed by a frontend application, the browser will not allow cross-origin HTTP requests. This means that resources can only be accessed from the exact origin as the API or origins permitted by the API.

FastAPI provides a CORS **middleware**, CORSMiddleware, that allows us to register domains which can access our API. The middleware takes an array of origins which will be permitted to access the resources on the server.

> **What is a middleware?**
>
> A middleware is a function that acts as an intermediary between an operation. In web APIs, a middleware serves as an mediator in a request-response operation.

For example, to allow only Packt to access our API, we define the URLs in the origin array:

```
origins = [
    "http://packtpub.com",
    "https://packtpub.com"
]
```

To allow requests from any client, the origins array will contain only one value, an asterisk (*). The asterisk is a wildcard that tells our API to allow requests from anywhere.

In main.py, lets configure our application to accept requests from everywhere:

```
from fastapi.middleware.cors import CORSMiddleware


# register origins

origins = ["*"]

app.add_middleware(
```

```
    CORSMiddleware,
    allow_origins=origins,
    allow_credentials=True,
    allow_methods=["*"],
    allow_headers=["*"],
)
```

In the code block above, we started by importing the `CORSMiddleware` class from FastAPI. We registered the origins array and finally registered the middleware into the application using the `add_middleware` method.

> **More information**
>
> The FastAPI documentation has more details on CORS - `https://fastapi.tiangolo.com/tutorial/cors/`

We have successfully configured our application to allow requests from any origin on the world wide web.

Summary

In this chapter, we learned how to secure a FastAPI application with OAuth and JWT. We also learned what dependency injection is, how it is used in FastAPI applications, and how to protect routes from unauthorized users. We also added a CORS middleware to permit access to our API from any client. We made use of the knowledge from previous chapters.

In the next chapter, you will be introduced to testing your FastAPI application. You will learn what testing an application is, why you should test applications, and how to test a FastAPI application.

Part 3: Testing And Deploying FastAPI Applications

On completing this part, you will be able to write and execute tests and deploy FastAPI applications using the knowledge obtained from the chapters included.

This part comprises the following chapters:

8
Testing FastAPI Applications

In the last chapter, we learned how to secure a FastAPI application using OAuth and **JSON Web Token (JWT)**. We successfully implemented an authentication system and learned what dependency injection is all about. We also learned how to inject dependencies into our routes to restrict unauthorized access and operations. We have successfully built a secure web API that has database support and is able to perform CRUD operations easily. In this chapter, we will learn what testing is and how to write tests to ensure that our application behaves as expected.

Testing is an integral part of the application development cycle. Application testing is done to ensure the correct functioning state of the application and easily detect anomalies in the application before deploying to production. Although we have been manually testing our application's endpoint in the last few chapters, we will be learning how to automate these tests.

By the end of this chapter, you will be able to write tests for your FastAPI application routes. This chapter will explain what unit testing is and how to perform unit testing on the application routes. In this chapter, you'll be covering the following topics:

- Unit testing with `pytest`
- Setting up our test environment

- Writing tests for REST API endpoints
- Test coverage

Technical requirements

For this chapter, you'll need a running MongoDB server on your local machine. The steps outlined in *Chapter 6, Connecting to a Database*, should be followed to get your database server up and running.

The code used in this chapter can be found at `https://github.com/ PacktPublishing/Building-Python-Web-APIs-with-FastAPI/tree/ main/ch08/planner`.

Unit testing with pytest

Unit testing is a testing procedure where individual components of an application are tested. This form of testing enables us to verify the working capability of individual components. For example, unit tests are employed in testing individual routes in an application to ensure the proper responses are returned.

In this chapter, we'll be making use of `pytest`, a Python testing library, to conduct our unit testing operations. Although Python comes with a unit testing library called `unittest` built in, the `pytest` library has a shorter syntax and is more preferred for testing applications. Let's install `pytest` and write our first sample test.

Let's install the `pytest` library:

```
(venv)$ pip install pytest
```

Next, create a folder called `tests` that will house the test files for our application:

```
(venv)$ mkdir tests && cd
(venv)$ touch __init__.py
```

Individual test filenames, during creation, will be prefixed with `test_`. This will enable the `pytest` library to recognize and run the test file. Let's create a test file in the newly created `tests` directory that checks the correctness of the addition, subtraction, multiplication, and division arithmetic operations:

```
(venv)$ touch test_arthmetic_operations.py
```

Let's define the function that performs the arithmetic operations first. In the `tests` file, add the following:

```
def add(a: int , b: int) -> int:
    return a + b


def subtract(a: int, b: int) -> int:
    return b - a


def multiply(a: int, b: int) -> int:
    return a * b


def divide(a: int, b: int) -> int:
    return b // a
```

Now that we have defined the operations to be tested, we'll create the functions that'll handle these tests. In the test functions, the operation to be executed is defined. The `assert` keyword is used to verify that the output on the left-hand side is in correspondence to the output of the operation on the right-hand side. In our case, we'll be testing that the arithmetic operations equal their respective results.

Add the following to the `tests` file:

```
def test_add() -> None:
    assert add(1, 1) == 2


def test_subtract() -> None:
    assert subtract(2, 5) == 3


def test_multiply() -> None:
    assert multiply(10, 10) == 100


def test_divide() -> None:
    assert divide(25, 100) == 4
```

> **Tip**
>
> The standard practice is to define the functions that will be tested in an external location (add(), subtract(), and so on, in our case). This file is then imported and the functions to be tested are invoked in the test functions.

With the test functions in place, we're set to run the test file. The tests can be executed by running the pytest command. However, this command runs all test files contained in the folder. To execute a single test, the test filename is passed as an argument. Let's run the test file:

```
(venv)$ pytest test_arithmetic_operations.py
```

The tests defined all passed. This is signified by the response in green:

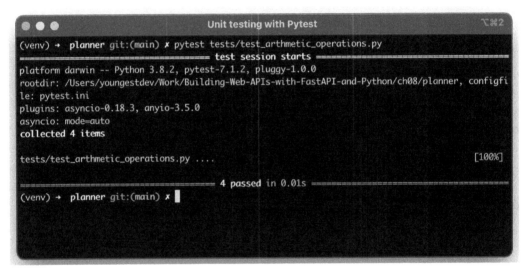

Figure 8.1 – Unit testing result carried out on arithmetic operations

Tests that failed, as well as the point of the failure, are highlighted in red. For example, say we modify the test_add() function as such:

```
def test_add() -> None:
    assert add(1, 1) == 11
```

In the following figure, the failing test, as well as the point of failure, is highlighted in red.

```
(venv) → tests git:(main) x
(venv) → tests git:(main) x pytest test_arthmetic_operations.py
================================ test session starts ================================
platform darwin -- Python 3.8.2, pytest-7.1.2, pluggy-1.0.0
rootdir: /Users/youngestdev/Work/Building-Web-APIs-with-FastAPI-and-Python/ch08/planner/tests, configfile: pytest.i
ni
plugins: asyncio-0.18.3, anyio-3.5.0, cov-3.0.0
asyncio: mode=auto
collected 4 items

test_arthmetic_operations.py F...                                            [100%]

===================================== FAILURES =====================================
_____ test_add _____

    def test_add():
>       assert add(1, 1) == 11
E       assert 2 == 11
E        +  where 2 = add(1, 1)

test_arthmetic_operations.py:18: AssertionError
============================= short test summary info ==============================
FAILED test_arthmetic_operations.py::test_add - assert 2 == 11
========================== 1 failed, 3 passed in 0.06s ===========================
(venv) → tests git:(main) x
```

Figure 8.2 – Failing test

The test failed at the `assert` statement, where the correct result, **2**, is displayed.

The failure is summarized as `AssertionError`, which tells us the test failed due to an incorrect assertion (**2 == 1**) being passed.

Now that we have an idea of how `pytest` works, let's take a look at fixtures in `pytest`.

Eliminating repetition with pytest fixtures

Fixtures are reusable functions defined to return the data needed in test functions. Fixtures are decorated with the `pytest.fixture` decorator. An example use case of a fixture is returning an applications instance to execute tests for the API endpoints. A fixture can be used to define the application client that is returned and used in test functions, eliminating the need to redefine the application instance in every test. We shall see how this is used in the *Writing tests for REST API endpoints* section.

Let's look at an example:

```
import pytest


from models.events import EventUpdate
```

```
# Fixture is defined.
@pytest.fixture
def event() -> EventUdpate:
    return EventUpdate(
        title="FastAPI Book Launch",
        image="https://packt.com/fastapi.png",
        description="We will be discussing the contents of
        the FastAPI book in this event.Ensure to come with
        your own copy to win gifts!",
        tags=["python", "fastapi", "book", "launch"],
        location="Google Meet"
    )


def test_event_name(event: EventUpdate) -> None:
    assert event.title == "FastAPI Book Launch"
```

In the preceding code block, we've defined a fixture that returns an instance of the EventUpdate pydantic model. This fixture is passed as an argument in the test_event_name function, enabling the properties to become accessible.

The fixture decorator can optionally take arguments. One of these arguments is scope – the scope of a fixture tells pytest what the duration of the fixture function will be. In this chapter, we'll make use of two scopes:

- session: This scope tells pytest to instantiate the function once for the whole testing session.
- module: This scope instructs pytest to execute the affixed function only once the test file is executed.

Now that we know what a fixture is, let's set up our test environment in the next section.

Setting up our test environment

In the previous section, we learned the basics of testing as well as what fixtures are. We will now test the endpoints for CRUD operations as well as user authentication. To test our asynchronous APIs, we'll be making use of httpx and installing the pytest-asyncio library to enable us to test our asynchronous API.

Install the additional libraries:

```
(venv)$ pip install httpx pytest-asyncio
```

Next, we'll create a configuration file called pytest.ini. Add the following code to it:

```
[pytest]
asyncio_mode = True
```

The configuration file is read when pytest is run. This automatically makes pytest run all tests in asynchronous mode.

With the configuration file in place, let's create the umbrella test file, conftest.py, which will be responsible for creating an instance of our application required by the test files. In the tests folder, create the conftest file:

```
(venv)$ touch tests/conftest.py
```

We'll start by importing the dependencies needed into conftest.py:

```
import asyncio
import httpx
import pytest


from main import app
from database.connection import Settings
from models.events import Event
from models.users import User
```

In the preceding code block, we have imported the asyncio, httpx, and pytest modules. The asyncio module will be used to create an active loop session to ensure the tests run on a single thread to avoid conflicts. The httpx test will act as the asynchronous client for conducting HTTP CRUD operations. The pytest library is needed for defining fixtures.

We have also imported our application's instance app, as well as the models and the Settings class. Let's define the loop session fixture:

```
@pytest.fixture(scope="session")
def event_loop():
    loop = asyncio.get_event_loop()
    yield loop
    loop.close()
```

With that in place, let's create a new database instance from the `Settings` class:

```
async def init_db():
    test_settings = Settings()
    test_settings.DATABASE_URL =
    "mongodb://localhost:27017/testdb"


    await test_settings.initialize_database()
```

In the preceding code block, we have defined a new `DATABASE_URL`, as well as invoking the initialization function defined in *Chapter 6, Connecting to a Database*. We're now making use of a new database, `testdb`.

Lastly, let's define the default client fixture, which returns an instance of our application run asynchronously through `httpx`:

```
@pytest.fixture(scope="session")
async def default_client():
    await init_db()
    async with httpx.AsyncClient(app=app,
    base_url="http://app") as client:
        yield client
        # Clean up resources
        await Event.find_all().delete()
        await User.find_all().delete()
```

In the preceding code block, the database is initialized first, and the application is spun as an `AsyncClient`, which is kept alive until the end of the test session. At the end of the testing session, the event and user collection are wiped off to ensure the database is empty before each test run.

In this section, you have been introduced to the steps involved in setting up your test environment. In the next section, you'll be taken through the process of writing tests for the each endpoint created in the application.

Writing tests for REST API endpoints

With everything in place, let's create the `test_login.py` file, where we'll test the authentication routes:

```
(venv)$ touch tests/test_login.py
```

In the test file, we'll start by importing the dependencies:

```
import httpx
import pytest
```

Testing the sign-up route

The first endpoint we'll be testing is the sign-up endpoint. We'll be adding the `pytest.mark.asyncio` decorator, which informs `pytest` to treat this as an async test. Let's define the function and the request payload:

```
@pytest.mark.asyncio
async def test_sign_new_user(default_client: httpx.AsyncClient)
-> None:
    payload = {
        "email": "testuser@packt.com",
        "password": "testpassword",
    }
```

Let's define the request header and expected response:

```
    headers = {
        "accept": "application/json",
        "Content-Type": "application/json"
    }


    test_response = {
        "message": "User created successfully"
    }
```

Now that we have defined the expected response for this request, let's initiate the request:

```
    response = await default_client.post("/user/signup",
json=payload, headers=headers)
```

Next, we'll test whether the request was successful by comparing the responses:

```
    assert response.status_code == 200
    assert response.json() == test_response
```

Before running this test, let's briefly comment out the line that erases user data in `conftest.py` as it will cause the authenticated tests to fail:

 # await User.find_all().delete()

From your terminal, start your MongoDB server and run the test:

```
(venv)$ pytest tests/test_login.py
```

The sign-up route has been successfully tested:

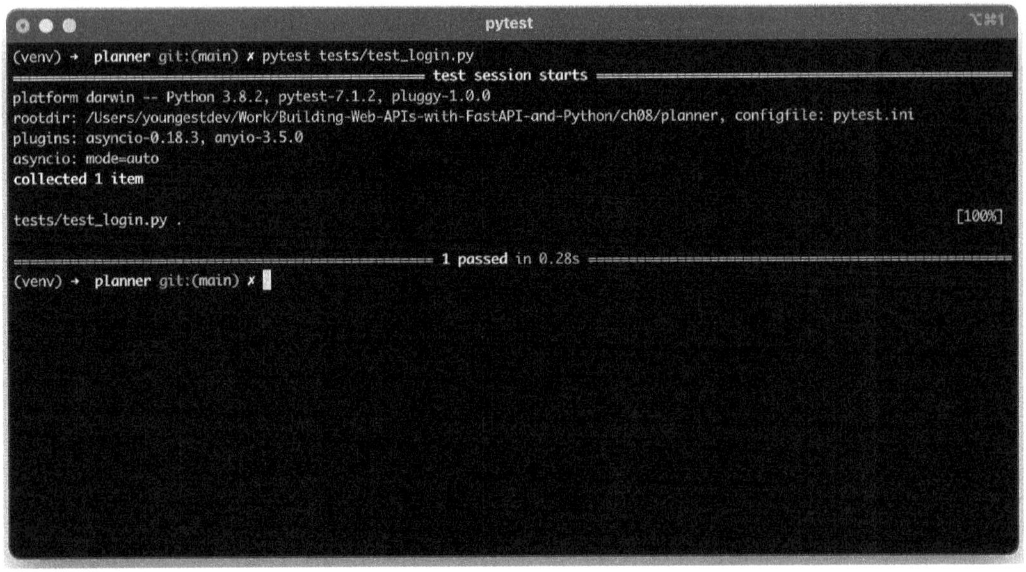

Figure 8.3 – Successful test run on the sign-up route

Let's proceed to write the test for the sign-in route. In the meantime, you can quickly tweak the test response to see whether your test fails or not!

Testing the sign-in route

Below the test for the sign-up route, let's define the test for the sign-in route. We'll start by defining the request payload and the headers before initiating the request like the first test:

```python
@pytest.mark.asyncio
async def test_sign_user_in(default_client: httpx.AsyncClient)
-> None:
    payload = {
        "username": "testuser@packt.com",
        "password": "testpassword"
```

```
    }

    headers = {
        "accept": "application/json",
        "Content-Type": "application/x-www-form-urlencoded"
    }
```

Next, we'll initiate the request and test the responses:

```
response = await default_client.post("/user/signin",
data=payload, headers=headers)

    assert response.status_code == 200
    assert response.json()["token_type"] == "Bearer"
```

Let's rerun the test:

```
(venv)$ pytest tests/test_login.py
```

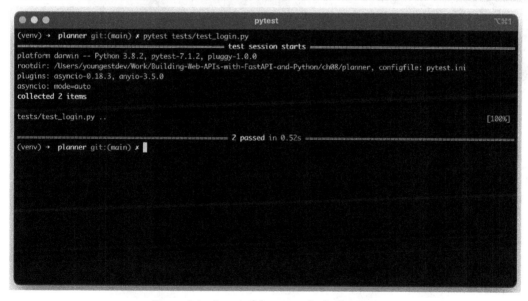

Figure 8.4 – Successful test run for both routes

Let's change the username for signing in to a wrong one to confirm that the test will fail:

```
payload = {
        "username": "wronguser@packt.com",
        "password": "testpassword"
    }
```

```
default_client = <httpx.AsyncClient object at 0x1044a84f0>

    @pytest.mark.asyncio
    async def test_sign_user_in(default_client: httpx.AsyncClient) -> None:
        payload = {
            "username": "wronguser@packt.com",
            "password": "testpassword"
        }

        headers = {
            "accept": "application/json",
            "Content-Type": "application/x-www-form-urlencoded"
        }

        response = await default_client.post("/user/signin", data=payload, headers=headers)

>       assert response.status_code == 200
E       assert 404 == 200
E        +  where 404 = <Response [404 Not Found]>.status_code

tests/test_login.py:41: AssertionError
========================= short test summary info =========================
FAILED tests/test_login.py::test_sign_user_in - assert 404 == 200
========================= 1 failed, 1 passed in 0.34s =========================
(venv) → planner git:(main) x █
```

Figure 8.5 – Failing test due to wrong request payload

We have successfully written tests for the sign-up and sign-in routes. Let's proceed to test the CRUD routes for the event planner API.

Testing CRUD endpoints

We'll start by creating a new file called test_routes.py:

```
(venv)$ touch test_routes.py
```

In the newly created file, add the following code:

```
import httpx
import pytest


from auth.jwt_handler import create_access_token
from models.events import Event
```

In the preceding code block, we have imported the regular dependencies. We've also imported the `create_access_token (user)` function and the `Event` model. Since some of the routes are protected, we'll be generating an access token ourselves. Let's create a new fixture that returns an access token when invoked. The fixture has a scope of `module`, which means it is run only once – when the test file is executed – and isn't invoked on every function call. Add the following code:

```
@pytest.fixture(scope="module")
async def access_token() -> str:
    return create_access_token("testuser@packt.com")
```

Let's create a new fixture that adds an event to the database. This action is performed to run preliminary tests before testing the CRUD endpoints. Add the following code:

```
@pytest.fixture(scope="module")
async def mock_event() -> Event:
    new_event = Event(
        creator="testuser@packt.com",
        title="FastAPI Book Launch",
        image="https://linktomyimage.com/image.png",
        description="We will be discussing the contents of
        the FastAPI book in this event.Ensure to come with
        your own copy to win gifts!",
        tags=["python", "fastapi", "book", "launch"],
        location="Google Meet"
    )


    await Event.insert_one(new_event)


    yield new_event
```

Testing READ endpoints

Next, let's write a test function that tests the **GET HTTP** method on the `/event` route:

```
@pytest.mark.asyncio
async def test_get_events(default_client: httpx.AsyncClient,
mock_event: Event) -> None:
    response = await default_client.get("/event/")
```

```
assert response.status_code == 200
assert response.json()[0]["_id"] == str(mock_event.id)
```

In the preceding code block, we're testing the event route path to check whether the event added to the database in the mock_event fixture is present. Let's run the test:

```
(venv)$ pytest tests/test_routes.py
```

Here's the result:

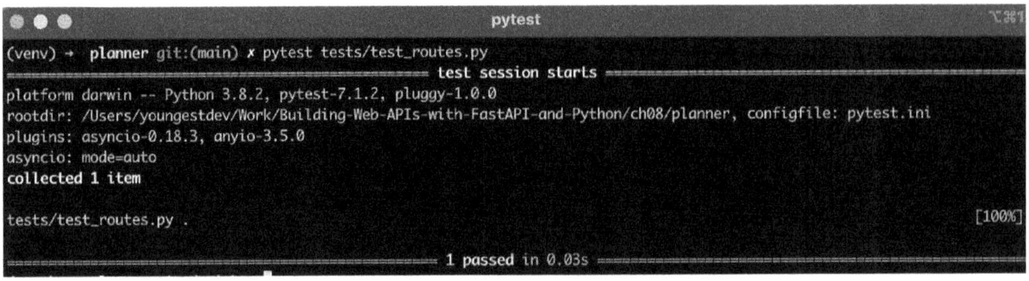

Figure 8.6 – Successful test run

Next, let's write the test function for the /event/{id} endpoint:

```
@pytest.mark.asyncio
async def test_get_event(default_client: httpx.AsyncClient,
mock_event: Event) -> None:
    url = f"/event/{str(mock_event.id)}"
    response = await default_client.get(url)

    assert response.status_code == 200
    assert response.json()["creator"] == mock_event.creator
    assert response.json()["_id"] == str(mock_event.id)
```

In the preceding code block, we're testing the endpoint that retrieves a single event. The event ID passed is retrieved from the mock_event fixture and the result from the request compared with the data stored in the mock_event fixture. Let's run the test:

```
(venv)$ pytest tests/test_routes.py
```

Here's the result:

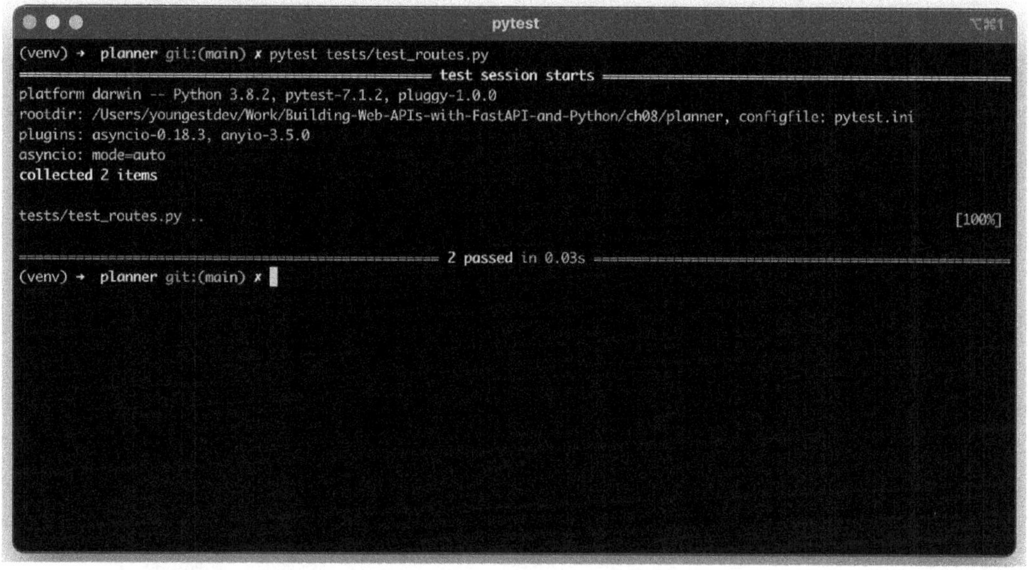

Figure 8.7 – Successful test run for single event retrieval

Next, let's write the test function for creating a new event.

Testing the CREATE endpoint

We'll start by defining the function and retrieving an access token from the fixture created earlier. We'll create the request payload, which will be sent to the server, the request headers, which will comprise the content type, as well as the authorization header value. The test response will also be defined, after which the request is initiated and the responses compared. Add the following code:

```
@pytest.mark.asyncio
async def test_post_event(default_client: httpx.AsyncClient,
access_token: str) -> None:
    payload = {
        "title": "FastAPI Book Launch",
        "image": "https://linktomyimage.com/image.png",
        "description": "We will be discussing the contents
        of the FastAPI book in this event.Ensure to come
        with your own copy to win gifts!",
        "tags": [
```

```
            "python",
            "fastapi",
            "book",
            "launch"
        ],
        "location": "Google Meet",
    }

    headers = {
        "Content-Type": "application/json",
        "Authorization": f"Bearer {access_token}"
    }

    test_response = {
        "message": "Event created successfully"
    }

    response = await default_client.post("/event/new",
    json=payload, headers=headers)

    assert response.status_code == 200
    assert response.json() == test_response
```

Let's rerun the test file:

```
(venv)$ pytest tests/test_routes.py
```

The result looks like so:

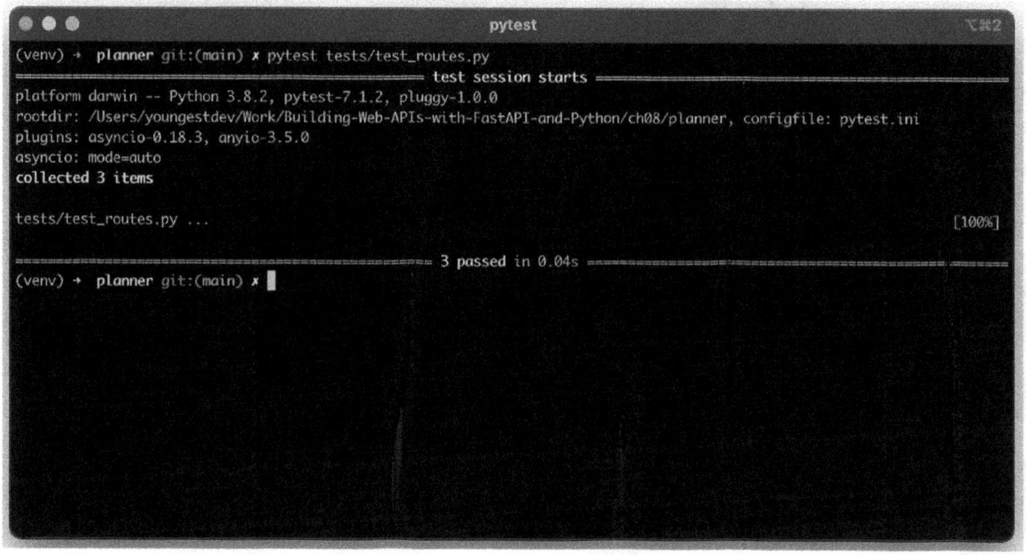

Figure 8.8 – Successful POST request test run

Let's write a test to verify the count of events stored in the database (in our case, 2). Add the following:

```
@pytest.mark.asyncio
async def test_get_events_count(default_client: httpx.
AsyncClient) -> None:
    response = await default_client.get("/event/")


    events = response.json()


    assert response.status_code == 200
    assert len(events) == 2
```

In the preceding code block, we have stored the JSON response in the events variable, whose length is used for our test comparison. Let's rerun the test file:

```
(venv)$ pytest tests/test_routes.py
```

Here's the result:

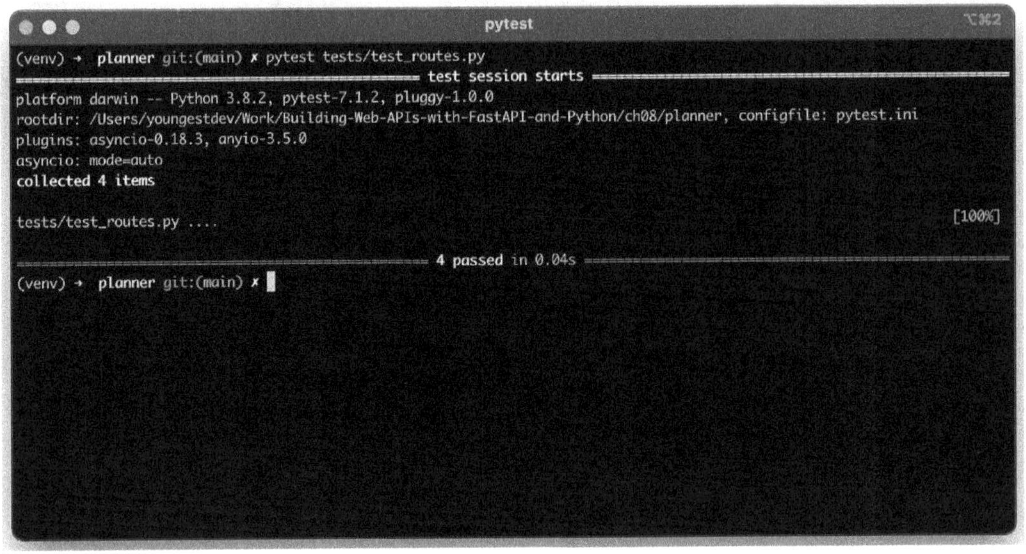

Figure 8.9 – Successful test run to confirm events count

We have successfully tested the GET endpoints /event and /event/{id} and the POST endpoint /event/new, respectively. Let's test the UPDATE and DELETE endpoints for /event/new next.

Testing the UPDATE endpoint

Let's start with the UPDATE endpoint:

```
@pytest.mark.asyncio
async def test_update_event(default_client: httpx.AsyncClient,
mock_event: Event, access_token: str) -> None:
    test_payload = {
        "title": "Updated FastAPI event"
    }

    headers = {
        "Content-Type": "application/json",
        "Authorization": f"Bearer {access_token}"
    }
```

```
url = f"/event/{str(mock_event.id)}"

response = await default_client.put(url,
json=test_payload, headers=headers)

assert response.status_code == 200
assert response.json()["title"] ==
test_payload["title"]
```

In the preceding code block, we are modifying the event stored in the database by retrieving the ID from the mock_event fixture. We then define the request payload and the headers. In the response variable, the request is initiated and the response retrieved is compared. Let's confirm that the test runs correctly:

```
(venv)$ pytest tests/test_routes.py
```

Here's the result:

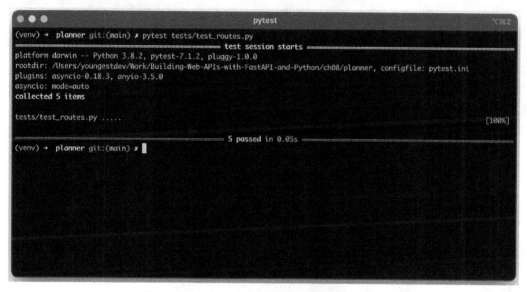

Figure 8.10 – Successful run for UPDATE request

> **Tip**
> The mock_event fixture comes in handy as the ID for MongoDB documents is uniquely generated every time a document is added to the database.

Let's change the expected response to confirm the validity of our test:

```
assert response.json()["title"] == "This test should
fail"
```

Rerun the test:

```
(venv) $ pytest tests/test_routes.py
```

Here's the result:

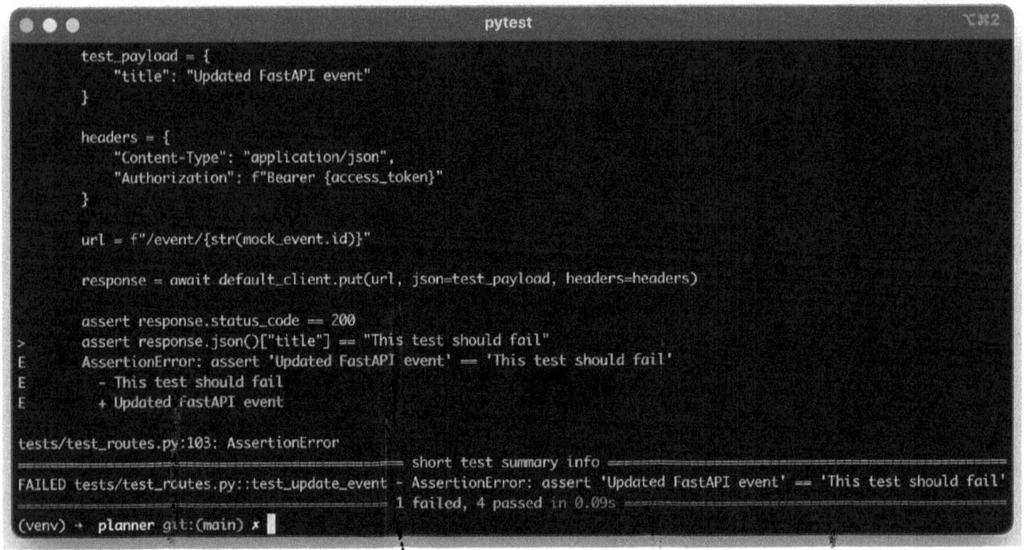

Figure 8.11 – Failed test due to difference in response objects

Testing the DELETE endpoint

Lastly, let's write the test function for the DELETE endpoint:

```
@pytest.mark.asyncio
async def test_delete_event(default_client: httpx.AsyncClient,
mock_event: Event, access_token: str) -> None:
    test_response = {
        "message": "Event deleted successfully."
    }

    headers = {
        "Content-Type": "application/json",
```

```
        "Authorization": f"Bearer {access_token}"
    }

    url = f"/event/{mock_event.id}"

    response = await default_client.delete(url,
    headers=headers)

    assert response.status_code == 200
    assert response.json() == test_response
```

Like the preceding tests, the expected test response is defined as well as the headers. The DELETE route is engaged and the response is compared. Let's run the test:

```
(venv)$ pytest tests/test_routes.py
```

Here's the result:

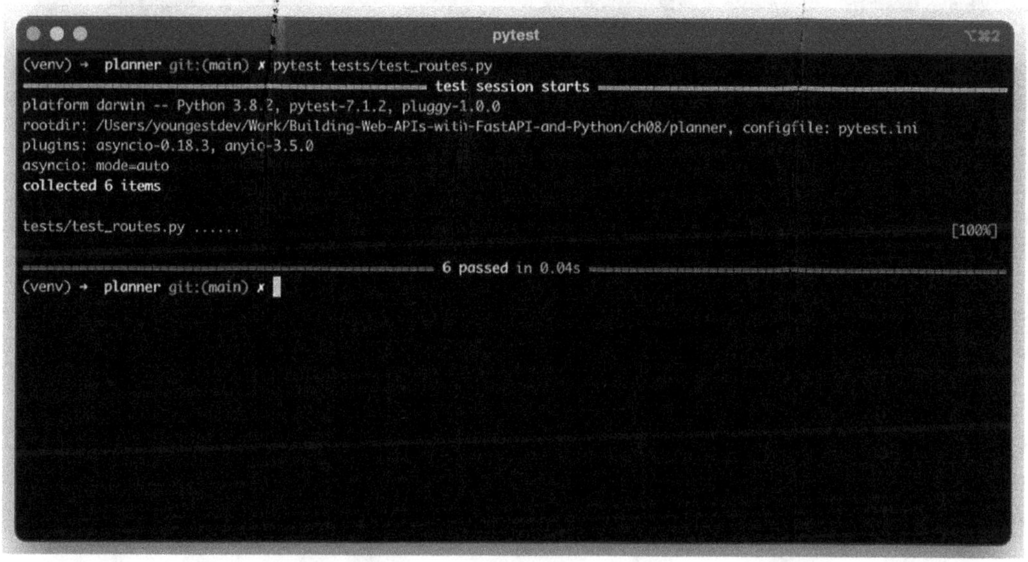

Figure 8.12 – Successful DELETE test

To confirm that the document has indeed been deleted, let's add a final test:

```
@pytest.mark.asyncio
async def test_get_event_again(default_client: httpx.
AsyncClient, mock_event: Event) -> None:
```

```
url = f"/event/{str(mock_event.id)}"
response = await default_client.get(url)

assert response.status_code == 200
assert response.json()["creator"] == mock_event.creator
assert response.json()["_id"] == str(mock_event.id)
```

The expected response is failure. Let's try it out:

```
(venv)$ pytest tests/test_routes.py
```

Here's the result:

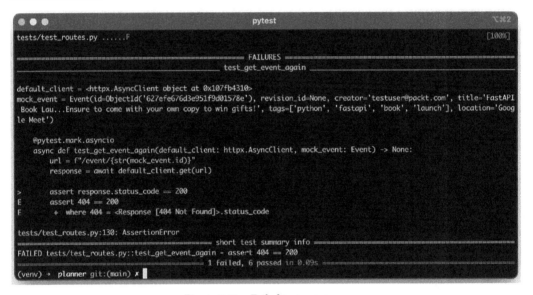

Figure 8.13 – Failed test response

As seen from the preceding screenshot, the item can no longer be found in the database. Now that you have successfully implemented the tests for authentication and event routes, uncomment the code responsible for clearing out user data from the database:

```
await User.find_all().delete()
```

Update the last test:

```
    assert response.status_code == 404
```

Lastly, let's run all the tests present in our application:

```
(venv)$ pytest
```

Here's the result:

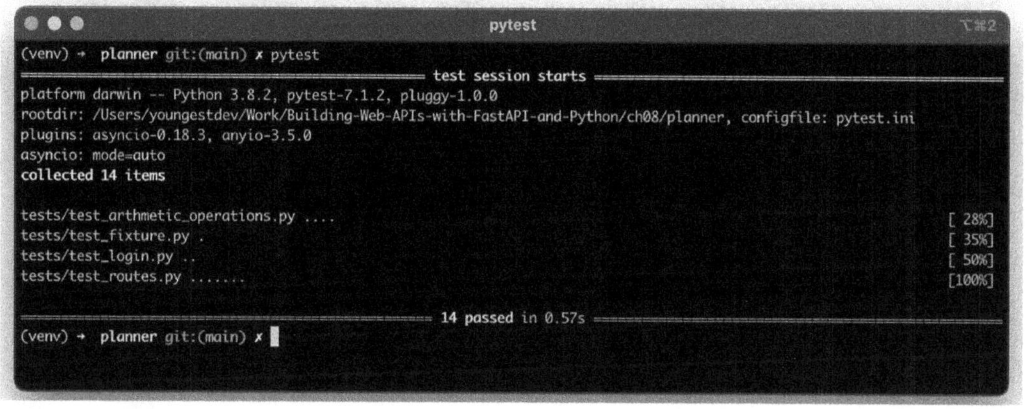

Figure 8.14 – Complete tests ran in 0.57 seconds

Now that we have successfully tested the endpoints contained in the event-planner API, let's run a coverage test to determine the percentage of our code involved in the test operation.

Test coverage

A test coverage report is useful in determining the percentage of our code that was executed in the course of testing. Let's install the `coverage` module so we can measure whether our API was adequately tested:

```
(venv)$ pip install coverage
```

Next, let's generate a coverage report by running this command:

```
(venv)$ coverage run -m pytest
```

Here's the result:

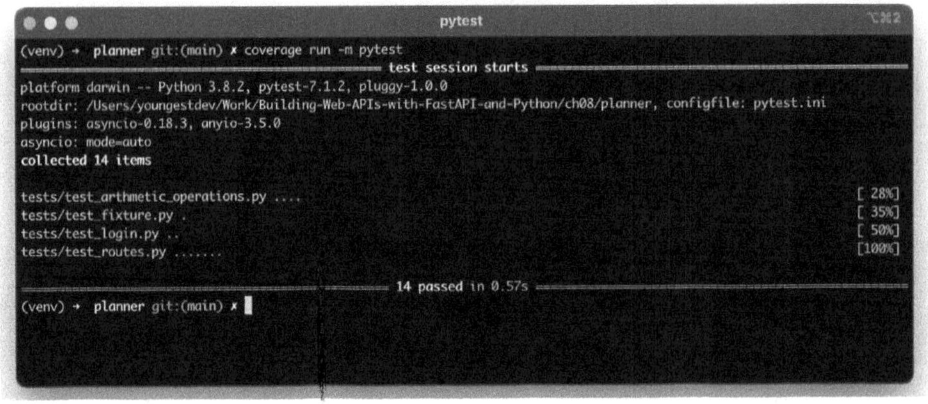

Figure 8.15 – Coverage report generated

Next, let's view the report generated by the `coverage run -m pytest` command. We can choose to view the report on the terminal or a web page by generating an HTML report. We'll do both.

Let's review the report from the terminal:

```
(venv)$ coverage report
```

Here's the result:

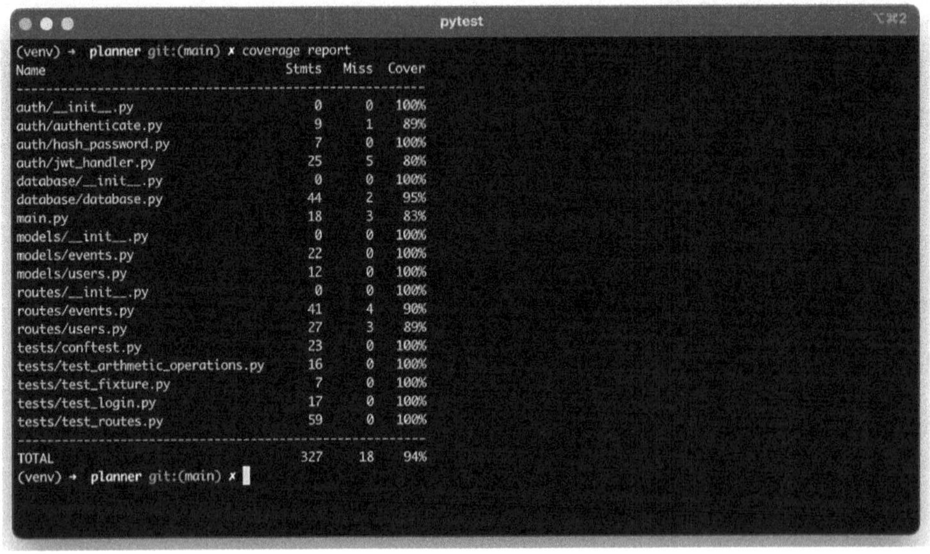

Figure 8.16 – Coverage report from the terminal

From the preceding report, the percentages signify the amount of code executed and interacted with. Let's generate the HTML report so we can check the blocks of code interacted with.

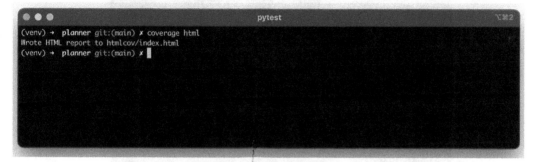

Figure 8.17 – Generating an HTML coverage report

Next, open `htmlcov/index.html` from your browser.

Module	statements	missing	excluded	coverage
auth/__init__.py	0	0	0	100%
auth/authenticate.py	9	1	0	89%
auth/hash_password.py	7	0	0	100%
auth/jwt_handler.py	25	5	0	80%
database/__init__.py	0	0	0	100%
database/connection.py	44	2	0	95%
main.py	21	3	0	86%
models/__init__.py	0	0	0	100%
models/events.py	22	0	0	100%
models/users.py	12	0	0	100%
routes/__init__.py	0	0	0	100%
routes/events.py	41	4	0	90%
routes/users.py	27	3	0	89%
tests/__init__.py	0	0	0	100%
tests/conftest.py	23	0	0	100%
tests/test_arthmetic_operations.py	16	0	0	100%
tests/test_fixture.py	7	0	0	100%
tests/test_login.py	17	0	0	100%
tests/test_routes.py	59	0	0	100%
Total	**330**	**18**	**0**	**95%**

Coverage report: 95%

coverage.py v6.3.3, created at 2022-07-12 17:02 +0100

Figure 8.18 – Coverage report from the web browser

Let's check the coverage report for `routes/events.py`. Click on it to display it.

```
Coverage for routes/events.py: 90%
41 statements    37 run    4 missing    0 excluded

1  from typing import List
2
3  from auth.authenticate import authenticate
4  from beanie import PydanticObjectId
5  from database.connection import Database
6  from fastapi import APIRouter, Depends, HTTPException, status
7  from models.events import Event, EventUpdate
8
9  event_router = APIRouter(
10     tags=["Events"]
11 )
12
13 event_database = Database(Event)
14
15
16 @event_router.get("/", response_model=List[Event])
17 async def retrieve_all_events() -> List[Event]:
18     events = await event_database.get_all()
19     return events
20
21
22 @event_router.get("/{id}", response_model=Event)
23 async def retrieve_event(id: PydanticObjectId) -> Event:
24     event = await event_database.get(id)
25     if not event:
26         raise HTTPException(
27             status_code=status.HTTP_404_NOT_FOUND,
28             detail="Event with supplied ID does not exist"
29         )
30     return event
31
32
33 @event_router.post("/new")
```

Figure 8.19 – Coverage report showing executed code in green and untouched code in red

Summary

In this chapter, you have successfully tested the API by writing tests for the authentication routes and the CRUD route. You have learned what testing is and how to write tests with `pytest`, a fast testing library built for Python applications. You also learned what `pytest` fixtures are and used them in creating reusable access tokens and database objects, as well as preserving the application instance throughout the testing session. You were able to assert the responses of your API HTTP requests and verify the behavior of your API. Finally, you learned how to generate a coverage report for your tests and distinguish the blocks of code run during the testing session.

Now that you have been equipped with the knowledge of testing web APIs, you are ready to publish your application to the World Wide Web through a deployment channel. In the next and final chapter, you'll learn how to containerize your application and deploy your locally using Docker and docker-compose.

9
Deploying FastAPI Applications

In the last chapter, you learned how to write tests for API endpoints created in a FastAPI application. We started by learning what testing means and walked through the basics of unit testing using the `pytest` library. We also looked at how to eliminate repetition and reuse test components with fixtures and then proceeded to set up our test environment. We wrapped up the last chapter by writing tests for each endpoint and then testing them alongside checking the test coverage reports after testing.

In this chapter, you'll learn how to deploy your FastAPI application locally using **Docker** and **docker-compose**. A brief section is also added with external resources to deploy your application on serverless platforms of your choice.

In this chapter, we'll be covering the following topics:

- Preparing for deployment
- Deploying with Docker
- Deploying Docker images

Technical requirements

The code used in this chapter can be found at `https://github.com/PacktPublishing/Building-Python-Web-APIs-with-FastAPI/tree/main/ch09/planner`.

Preparing for deployment

Deployment usually marks the end of an application's life cycle. Before deploying our applications, we must make sure the right settings required for a smooth deployment are put in place. These settings include ensuring the application dependencies are up to date in the `requirements.txt` file, configuring environment variables, and so on.

Managing dependencies

In a few earlier chapters, we installed packages such as `beanie` and `pytest`. These packages are absent from the `requirements.txt` file, which serves as the dependency manager for our application. It is important that the `requirements.txt` file is kept up to date.

In Python, the list of packages used in a development environment can be retrieved using the `pip freeze` command. The `pip freeze` command returns a list of all packages installed directly and the sub-dependencies for each package installed. Luckily, the `requirements.txt` file can be maintained manually, enabling us to list only the main packages, thereby making dependency management easier.

Let's list the dependencies used by the application before overwriting the `requirements.txt` file:

```
(venv)$ pip freeze
anyio==3.5.0
asgi-lifespan==1.0.1
asgiref==3.5.0
attrs==21.4.0
bcrypt==3.2.2
cffi==1.15.0
python-multipart==0.0.5
...
```

The command returns several dependencies, some of which we do not use directly in the application. Let's manually fill the `requirements.txt` file with the packages we will be using:

requirements.txt

```
fastapi==0.78.0
bcrypt==3.2.2
beanie==1.11.1
email-validator==1.2.1
httpx==0.22.0
Jinja2==3.0.3
motor==2.5.1
passlib==1.7.4
pytest==7.1.2
python-multipart==.0.0.5
python-dotenv==0.20.0
python-jose==3.3.0
sqlmodel==0.0.6
uvicorn==0.17.6
```

In this code block, we have populated the `requirements.txt` file with the dependencies used directly in our application.

Configuring environment variables

We used environment variables in *Chapter 6, Connecting to a Database*. Environment variables can be injected during deployment, as we'll see in the next section.

> **Note**
> It is important to note that environment variables are to be properly handled and kept out of version control systems such as GitHub.

Now that we have covered the necessary steps in preparation for our deployments, let's proceed to deploying our application locally with Docker in the next section.

Deploying with Docker

In *Chapter 1*, *Getting Started with FastAPI*, you were introduced to the basics of Docker and the Dockerfile. In this section, you'll be writing a Dockerfile for the event planner API.

Docker is the most popular technology used for containerization. Containers are self-contained systems consisting of packages, code, and dependencies that enable them to run in different environments with little to no dependence on their running environment. Docker uses Dockerfiles for the containerization process.

Docker can be used for local development as well as for deploying applications to production. We'll only be looking at local deployment in this chapter, and links to official guides on deploying to cloud services will be included as well.

For managing applications with multiple containers, such as an application container and a database container, the compose tool is used. Compose is a tool used to manage multi-container Docker applications defined in the configuration file, usually `docker-compose.yaml`. The compose tool, docker-compose, comes installed with the Docker engine.

Writing the Dockerfile

A Dockerfile contains a set of instructions employed to build a Docker image. The Docker image built can then be distributed to registries (private and public), deployed to cloud servers such as AWS and Google Cloud, and used on different operating systems by creating a container.

Now that we know what a Dockerfile does, let's create a Dockerfile to build the application's image. In the project directory, create the `Dockerfile` file:

```
(venv)$ touch Dockerfile
```

Dockerfile

```
FROM python:3.10

WORKDIR /app

COPY requirements.txt /app

RUN pip install --upgrade pip && pip install -r /app/
requirements.txt
```

```
EXPOSE 8080

COPY ./ /app

CMD ["python", "main.py"]
```

Let's go through the instructions contained in the preceding Dockerfile one by one:

- The first instruction the Dockerfile executes is to set the base image for our own image using the FROM keyword. Other variations of this image can be found at https://hub.docker.com/_/python.

- The next line uses the WORKDIR keyword to set the working directory to /app. A working directory helps organize the structure of the project being built to an image.

- Next, we copy the requirements.txt file from the local directory to the working directory on the Docker container using the COPY keyword.

- The next instruction is the RUN command, which is used to upgrade the pip package and then install the dependencies from the requirements.txt file.

- The next command exposes the PORT from which our application can be accessed from the local network.

- The next command copies the rest of the files and folders into the Docker container working directory.

- Lastly, the last command starts the application using the CMD command.

Each set of instructions listed in the Dockerfile is built as an individual layer. Docker does an intelligent job of caching each layer during a build to reduce build time and eliminate repetition. If a layer that is essentially an instruction is untouched, the layer is skipped and the previously built one is used. That is, Docker uses the cache system when building images.

Let's create a .dockerignore file before proceeding to build our image:

```
(venv)$ touch .dockerignore
```

.dockerignore

```
Venv
.env
.git
```

> **What Is .dockerignore?**
>
> The .dockerignore file contains files and folders to be exempted from instructions defined in the Dockerfile.

Building the Docker image

To build the application image, run the following command in the base directory:

```
(venv)$ docker build -t event-planner-api .
```

This command simply tells Docker to build an image with the event-planner-api tag from the instructions defined in the current directory, which is represented by the dot at the end of the command. The build process commences once the command is run and the instructions are executed:

Figure 9.1 – Docker build process

Now that we have successfully built our application's image, let's pull the MongoDB image:

```
(venv)$ docker pull mongo
```

We're pulling a MongoDB image to create a standalone database container accessible from the API container when created. By default, the Docker container has a separate network configuration, and connecting to the machine hosts' localhost address is not allowed.

Figure 9.2 – Pulling a MongoDB image

> **What Is docker pull?**
>
> The `docker pull` command is responsible for downloading images from a registry. Unless otherwise stated, these images are downloaded from the public Docker Hub registry.

Deploying our application locally

Now that we have created the images for the API and pulled the image for the MongoDB database, let's proceed to write a compose manifest to handle our application deployment. The docker-compose manifest will consist of the API service and the MongoDB database service. In the root directory, create the manifest file:

```
(venv)$ touch docker-compose.yml
```

The contents of the docker-compose manifest file will be as follows:

docker-compose.yml

```yaml
version: "3"

services:
  api:
    build: .
    image: event-planner-api:latest
    ports:
      - "8080:8080"
    env_file:
      - .env.prod

  database:
    image: mongo
    ports:
      - "27017"
    volumes:
      - data:/data/db

  volumes:
    data:
```

In the `services` section, we have the `api` service and the `database` service. In the `api` service, the following set of instructions are put in place:

- The `build` field instructs Docker to build the `event-planner-api:latest` image for the `api` service from the Dockerfile situated in the current directory denoted by `.`.

- Port `8080` is exposed from the container to enable us to access the service through HTTP.

- The environment file is set to `.env.prod`. Alternatively, the environment variables can be set in this format:

```
environment:
    - DATABASE_URL=mongodb://database:27017/planner
    - SECRET_KEY=secretkey
```

This format is mostly used when environment variables are to be injected from a deployment service. It is encouraged to use the environment file.

In the database service, the following set of instructions are put in place:

- The `database` service makes use of the `mongo` image we pulled earlier.

- Port `27017` is defined but not exposed externally. The port is only accessible internally by the `api` service.

- A persistent volume is attached to the service to store our data. The folder allocated for this is `/data/db`.

- Lastly, the volume for this deployment is created with the name `data`.

Now that we have understood the content of the compose manifest, let's create the environment file, `.env.prod`:

.env.prod

```
DATABASE_URL=mongodb://database:27017/planner
SECRET_KEY=NOTSTRONGENOUGH!
```

In the environment file, `DATABASE_URL` is set to the name of the MongoDB service created by the compose manifest.

Running our application

We are set to deploy and run the application from the docker-compose manifest. Let's start the services using the compose tool:

```
(venv)$ docker-compose up -d
```

This command starts the services in detached mode:

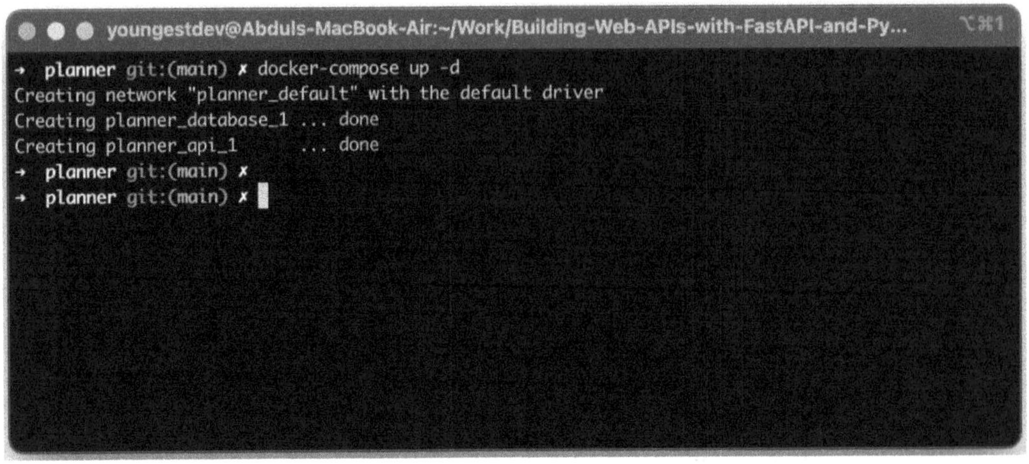

Figure 9.3 – Starting our application using the docker-compose tool

The application services have been created and deployed. Let's verify by checking the list of containers running:

```
(venv)$ docker ps
```

The result is as follows:

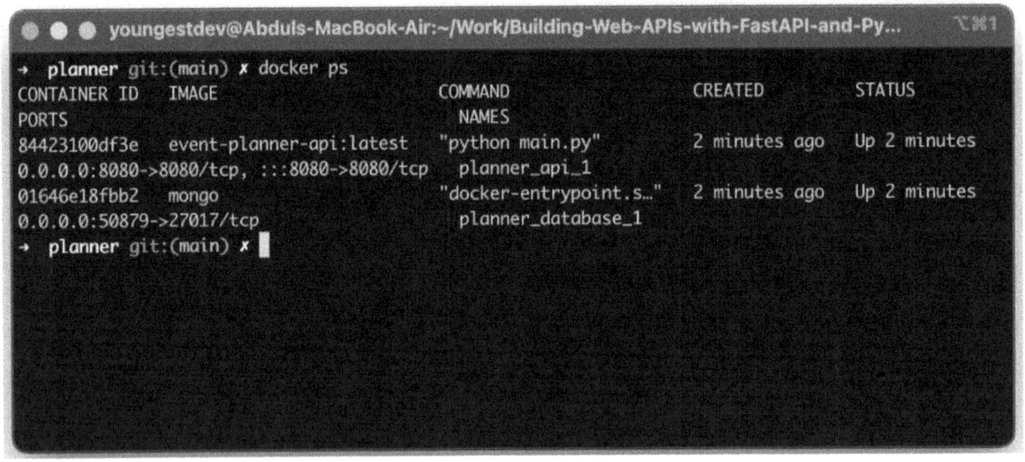

Figure 9.4 – List of running containers

The command returns the list of containers running alongside ports in which they can be accessed. Let's test the working state by sending a GET request to the deployed application:

```
(venv)$ curl -X 'GET' \
  'http://localhost:8080/event/' \
  -H 'accept: application/json'
```

We get the following response:

```
[]
```

Great! The deployed application works correctly. Let's verify that the database works as well by creating a user:

```
(venv)$ curl -X 'POST' \
  'http://localhost:8080/user/signup' \
  -H 'accept: application/json' \
  -H 'Content-Type: application/json' \
  -d '{
  "email": "fastapi@packt.com",
  "password": "strong!!!"
}'
```

We get a positive response as well:

```
{
  "message": "User created successfully"
}
```

Now that we have tested both routes, you can go ahead to test the other routes. To stop the deployment server after exploring, the following command is run from the root directory:

```
(venv)$ docker-compose down
```

The result is as follows:

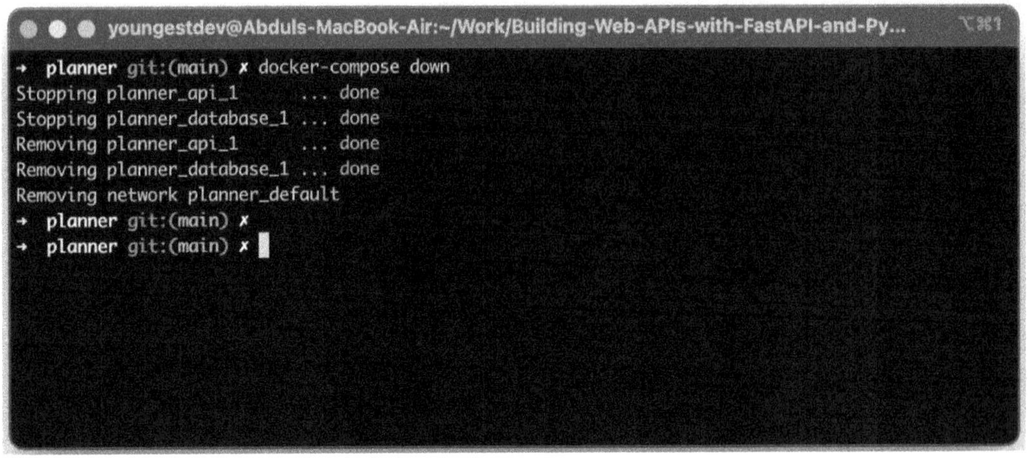

Figure 9.5 – Stopping application instances

Deploying Docker images

In the last section, we learned how to build and deploy Docker images locally. These images can be deployed on any virtual machine and on serverless platforms such as Google Cloud and AWS.

The regular mode of operation involves pushing your Docker images to a private registry on the serverless platform. The process involved in deploying Docker images on serverless platforms varies from provider to provider and, as a result, the links to select serverless service providers have been provided here:

- Google Cloud Run: `https://cloud.google.com/run/docs/quickstarts/build-and-deploy/python`

- Amazon EC2: `https://docs.aws.amazon.com/AmazonECS/latest/developerguide/getting-started-ecs-ec2.html`

- Deploying to Microsoft Azure: `https://docs.microsoft.com/en-us/azure/container-instances/container-instances-tutorial-deploy-app`

The steps covered in the previous section can be followed when the Docker images are to be installed on a machine or traditional server.

Deploying databases

Platforms such as Google Cloud, AWS provide the option to host your database containers. However, this might be expensive in terms of running cost and overall manageability.

For platforms that do not support the deployment of docker-compose manifests, the MongoDB database can be hosted on MongoDB Atlas (`https://www.mongodb.com/atlas/database`) and the `DATABASE_URL` environment variable overwritten with the connection string. A detailed guide on setting up a database on MongoDB atlas can be found at `https://www.mongodb.com/docs/atlas/getting-started/`.

Summary

In this chapter, you learned how to prepare your application for deployment. You started by updating the dependencies used for the application and saving them in the `requirements.txt` file before moving on to managing the environment variables used for the API.

We also covered the steps involved in deploying an application to production: building the Docker image from the Dockerfile, configuring the compose manifest for the API and database services, and then deploying the application. You also learned new commands to check the list of running containers as well as start and stop Docker containers. Finally, you tested the application to ensure the deployment was successful.

This marks the end of this book, and you should now be ready to build, test, and deploy a FastAPI application on the web. We covered various concepts and ensured that each concept is properly discussed with adequate examples: routing, templating, authentication, connecting to the database, and deploying the application. Be sure to check out the external resources mentioned occasionally in the book to gain more knowledge!

Index

Symbols

A

B

C

D

Packt.com

Subscribe to our online digital library for full access to over 7,000 books and videos, as well as industry leading tools to help you plan your personal development and advance your career. For more information, please visit our website.

Why subscribe?

- Spend less time learning and more time coding with practical eBooks and Videos from over 4,000 industry professionals

- Improve your learning with Skill Plans built especially for you

- Get a free eBook or video every month

- Fully searchable for easy access to vital information

- Copy and paste, print, and bookmark content

Did you know that Packt offers eBook versions of every book published, with PDF and ePub files available? You can upgrade to the eBook version at packt.com and as a print book customer, you are entitled to a discount on the eBook copy. Get in touch with us at customercare@packtpub.com for more details.

At www.packt.com, you can also read a collection of free technical articles, sign up for a range of free newsletters, and receive exclusive discounts and offers on Packt books and eBooks.

Other Books You May Enjoy

If you enjoyed this book, you may be interested in these other books by Packt:

Python Web Development with Sanic

Adam Hopkins

ISBN: 978-1-80181-441-6

- Understand the difference between WSGI, Async, and ASGI servers
- Discover how Sanic organizes incoming data, why it does it, and how to make the most of it
- Implement best practices for building reliable, performant, and secure web apps
- Explore useful techniques for successfully testing and deploying a Sanic web app

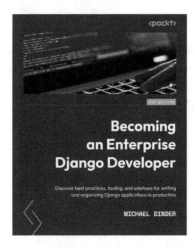

Becoming an Enterprise Django Developer

Michael Dinder

ISBN: 978-1-80107-363-9

- Create and configure a proof-of-concept Django project on a local machine
- Understand the steps and tools used to scale up a proof-of-concept project to production without going too deep into specific technologies
- Explore core Django components and how to use them in different ways to suit your app's needs
- Find out how Django allows you to build RESTful APIs
- Write and run test cases using the built-in testing tools in Django

Packt is searching for authors like you

If you're interested in becoming an author for Packt, please visit authors. packtpub.com and apply today. We have worked with thousands of developers and tech professionals, just like you, to help them share their insight with the global tech community. You can make a general application, apply for a specific hot topic that we are recruiting an author for, or submit your own idea.

Share Your Thoughts

Now you've finished *Building Python Web APIs with FastAPI*, we'd love to hear your thoughts! Scan the QR code below to go straight to the Amazon review page for this book and share your feedback or leave a review on the site that you purchased it from.

https://packt.link/r/1801076634

Your review is important to us and the tech community and will help us make sure we're delivering excellent quality content.